I0813622

A DARK HISTORY *of* WHISKY

Sharing a whisky with friends is undoubtably the best mixer there is and this book is dedicated in the spirit of friendship, respect and gratitude to Richie Williams by way of an apology for the look of sheer horror upon his face when I drank his Dalmore, and Anthony Brannan, whom I see far less than I would like, but will always be a positive in my life.

Cheers, guys!
Gary M. Dobbs

A DARK HISTORY *of* WHISKY

GARY DOBBS

AN IMPRINT OF PEN & SWORD BOOKS LTD.
YORKSHIRE – PHILADELPHIA

First published in Great Britain in 2025 by
PEN AND SWORD HISTORY
An imprint of
Pen & Sword Books Ltd
Yorkshire – Philadelphia

ISBN 978 1 39903 406 7

A CIP catalogue record for this book is available from the British Library.

Typeset in Times New Roman 12/16 by
SJmagic DESIGN SERVICES, India.
Printed and bound in the UK by CPI Group (UK) Ltd.

Pen & Sword Books Limited incorporates the imprints of Archaeology, Atlas, Aviation, Battleground, Digital, Discovery, Family History, Fiction, History, Local, Local History, Maritime, Military, Military Classics, Politics, Select, Transport, True Crime, After the Battle, Air World, Claymore Press, Frontline Publishing, Leo Cooper, Remember When, Seaforth Publishing, The Praetorian Press, Wharncliffe Books, Wharncliffe Local History, Wharncliffe Transport, Wharncliffe True Crime and White Owl.

For a complete list of Pen & Sword titles please contact
PEN & SWORD BOOKS LIMITED
George House, Units 12 & 13, Beevor Street, Off Pontefract Road,
Barnsley, South Yorkshire, S71 1HN, England
E-mail: enquiries@pen-and-sword.co.uk
Website: www.pen-and-sword.co.uk

or

PEN AND SWORD BOOKS
1950 Lawrence Rd, Havertown, PA 19083, USA
E-mail: uspen-and-sword@casematepublishers.com
Website: www.penandswordbooks.com

Contents

Introduction vi

Chapter 1 Distillation and the Birth of Modern Whisky 1
Chapter 2 By the Light of the Silvery Moon 16
Chapter 3 Trouble on the Emerald Isle 30
Chapter 4 A Whisky Mixer 42
Chapter 5 The American Experience 62
Chapter 6 Old Glory, Independence and Whiskey 67
Chapter 7 The Birth of Bourbon 80
Chapter 8 Whisky in Pop Culture 85
Chapter 9 Prohibition and the Rise of the American Gangster 95
Chapter 10 The Wide World of Whisky 113
Chapter 11 How Whisky is Made 128
Chapter 12 A Double Whisky Mixer 138
Chapter 13 How to Taste Whisky 160
Chapter 14 Let's Taste Whisky and Throw in Some Mixers 166

Glossary 186
Suggested Reading 189
Index 190

Introduction

Is it whisky or whiskey?

Confusingly it is both and there is no real definitional difference between whisky or whiskey; they are both whisky with or without the 'e'. It all comes from the way the word whisky or whiskey has evolved over the centuries and stems from the original Gaelic form of the word, but alas even that is confusing. In Irish Gaelic the word whiskey evolved from the phrase *uisce beatha*, which means the water of life, but in Scots Gaelic, a language with great similarities to the Irish, the phrase is spelled *uisge beatha*, which also means water of life. Both mean the same and yet the spellings are different. It is widely believed that the word whisky or whiskey is a corruption of the original Gaelic and that the phrase has been progressively anglicised to uiskie, and then usky until it came to be whisky or whiskey. It's all down to phonetics and the word *uisge/uisce* is pronounced oos-key and so, drunkenly, it becomes whisky or, wait for it, whiskey.

Simple enough on the face of it, but yet if one delves a little deeper into the subject, further confusion is added. For instance, in 1785 no less a personage than Robert Burns in his piece *Earnest Cry and Prayer* dropped the original Gaelic and instead used the Latin phrase for the water of life, and it became aqua vitae.

Tell them wha hae the chief direction,
Scotland an' me's in great affliction
E'er sin' they laid that curst restriction
On Aqua Vitae.

In fact, the first written reference to aqua vitae comes when in 1494 King James IV sent instructions to Friar John Cor to make aqua vitae from the eight bolls of malt the crown had provided. A boll being an archaic measurement that equals 24 pecks. It doesn't help that the word peck is now archaic in itself – however, it basically equates to 2 gallons of dry goods, usually a foodstuff. What is certain is that a year before his boozy request, King James IV was campaigning against the Lords of the Isles on the west coast of Scotland and that he first encountered aqua vitae there, where it will have been known as *uisge beatha*.

Now to go off on a tangent, we can extrapolate that those 8 bolls of malt are equal to roughly 1,024 dry gallons of the stuff, assuming, of course, that it was well compacted. And so, with the weight of each dry gallon coming in at approximately 8lb, then we come up somewhere around 8,000lb of malt. This means that the order would have been sufficient to make somewhere in the region of 790 gallons of whisky/whiskey/aqua vitae, *uisge beatha*/*uisce beatha*. This mathematical lunge depends on the fact that the final product was watered down to a standard 80% proof. The reader will realise that we can't be certain of these calculations, but for the sake of this narrative it is as good a hypothesis as any other.

So, there we have it, only we are no closer to understanding why whisky or whiskey is spelled two different ways. Perhaps, an easier way to address the problem is to accept that in Scotland, Canada and Japan it is whisky, while in Ireland, America and mostly but not everywhere else it is whiskey. Only it isn't and certain American brands such as Maker's Mark perversely use the Scotch spelling of whisky without the e. Sometimes it seems a free for all regarding the spelling.

Throughout this book, I will use both spellings. If a certain brand uses the e, then I will use the e and if it doesn't then I won't.

To E or not to E, indeed.

At this point I should point out that the book you hold in your hands is not a standard whisky book. Of course, it does contain

Pattisons, not an 'e' in sight. *Author's collection*

information that can be found in countless other whisky books such as tasting notes, distillery histories and the general development of the golden liquid we all adore. A whisky book that did not contain such information would be odd indeed, but the emphasis in the following narrative is very much on the darker history of the drink. Whisky noir, if you please. The text is also interspaced with whisky quotes, some well known, others obscure, so the reader will never be without an interesting snippet of conversation when sharing a whisky with friends.

So why not make yourself comfortable, kick off your shoes, pour yourself a dram and settle down. Now take my hand as I lead you through a dark history of the wonderful golden liquid, we call whisky/whiskey/*uisce beatha*/*uisge beatha* or simply aqua vitae.

For it truly is the water of life.

Chapter 1

Distillation and the Birth of Modern Whisky

Whisky in terms of popularity has never been in such rude health. Only a few decades ago those in search of a new whisky experience would have been severely hampered by the limited range generally available. Supermarkets and off-licences tended to stock only a core range of blended whiskies and, if you were lucky, a handful of single malts. All that has changed and with the advent of online shopping consumers can browse endless virtual shelves bursting with both the commonplace and the exotic; a single click of a mouse can send a familiar delight or a new discovery whirling its way to your home. There are whisky clubs in which for the price of a monthly subscription you will be sent several different whisky samples each month that can be geared to your own tastes depending on the flavour profile you indicated. There are whisky societies with meetings held right across the country from the furthest reaches of the Highlands to maybe the public house on your own street. There are scores of YouTube channels dedicated to whisky tasting, websites featuring both whisky reviews and news and even social media groups devoted to the wonderfully complex drink that is whisky. Though beware, for many of the online whisky reviewers will try to convince you that single malts are superior to blends, and that expensive is better, when this is far too simplistic, and, to be perfectly blunt, nonsense.

Great fury, like great whisky requires long fermentation.

Truman Capote

Whisky is a classy drink and carries an air of sophistication. It doesn't matter if one prefers Scotch or bourbon, Irish, rye, Japanese, Welsh or even Indian as their tipple of choice. Whisky is whisky, that is unless it's whiskey of course.

When you take your nightly dram, know that you are in good company for, as every whisky lover knows, there are few pleasures as rewarding as ending a long day with a dram of that golden liquid. Think of Frank Sinatra, who was buried with a bottle of his beloved Jack Daniel's Old No. 7, or Mark Twain, who famously declared that, 'Too much of a good thing is bad for one, but too much good whiskey is barely enough.' The author Raymond Chandler may have preferred gin as his tipple of choice but his creation, Philip Marlowe, the creation being far cooler, hipper and downright appealing than the creator, that man without tarnish, was always knocking back a slug of whiskey. Winston Churchill was a whisky aficionado who claimed he got the taste during the South African War when it was safer than drinking the water, and the author George Bernard Shaw aptly called whisky 'liquid sunshine'.

Bring me sunshine, indeed.

Whisky also spreads a large footprint across popular culture and several of the most iconic TV shows of recent years have given the drink a new prominence, a touch of coolness that has blasted away the once-held image of it being an old man's drink. Walter White often indulged in Dimple Scotch whisky in *Breaking Bad*, Don Draper in *Mad Men* was rarely seen without a bottle of Canadian Club blended whisky, barely an episode of *Yellowstone* went by without John Dutton or one or other of his dysfunctional family moodily sipping a dram of bourbon, and the gangster show *Peaky Blinders* helped propel the sales of Irish whiskey. In fact, the Peaky Blinders even have a whiskey, Irish of course, named after them. The same can be said of the big screen and even James Bond, that lover of vodka martinis, is seen drinking a dram or two of The Macallan in *Skyfall*. *Blade Runner* uses a bottle of Johnnie Walker as a futuristic prop, Batman's butler Alfred seems to enjoy a dram of Talisker, and

Liquid sunshine. *Courtesy Cotswold Distillery*

Woody Harrelson in *Zombieland* favours Dewar's whisky, when he's not decapitating zombies that is.

Whisky is no longer, indeed if it ever was, a drink solely produced in Scotland and Ireland. All over the world the water of life is being produced. In England, Wales, America, Canada, Germany, Taiwan, Japan, Finland and Australia the amber liquid flows from the stills. At the time of writing worldwide whisky revenue amounted to US $91.70 billion and was expected to grow annually by 7.36% between 2024 and 2026.

The whisky industry is booming and the future seems incredibly bright, but whisky has a long, often dark, history behind it and it is that history that we shall explore throughout the pages of this book. I've already stated that those looking for tasting notes need not fret, you'll find those towards the latter section of the book, but for now let's go back to the very birth of this fascinating and complex drink. I hope you; dear reader, are holding a dram as you turn the pages because I know I am as I write them.

> *What whiskey will not cure, there is no cure for.*
>
> Old Irish proverb

So, who was it? The genius who first decided to take beer and boil it in a still?

Some claim it was a Celtic shaman, while others say it was a man of the cloth. It's largely conjecture, though we do know that the origins of distillation stretch back through time. All the way back, in fact, to Persia and the works of Al Jabir (AD 721–841), but even that is disputed and archaeologists have found equipment believed to have been used for distillation that has been dated far back into pre-history. There is no indication that these early forms of distillation were used to create an alcoholic drink but were more likely used to fashion liquids that would have served in cosmetics and medicines. Distillation is simply the process of concentration or separation of a liquid via the process of evaporation and condensation, and although the first real use of distilling technology is hotly debated by historians, most agree that there is enough evidence to say that primitive distillation occurred before 2000 BC.

It was around AD 200 that the peripatetic philosopher Alexander of Aphrodisias wrote of a process of distilling seawater into drinking water. Ships were limited to the amount of water they could carry and so seawater would be collected in buckets and then poured into a boiling device. The resulting steam would be collected, which created perfectly distilled fresh water. Later, during the third century, we have actual documentation of a more purposed distillation system being used in Egypt. The Egyptian-born Greek alchemist Zosimos of Panopolis described a process in which tubes were connected to vessels heated by a flame. The liquids would evaporate, travel down the tubes and condense in a collection vessel. This description is very close to the basic way a modern still operates. However, it was during the ninth century that the aforementioned Al Jabir is credited with inventing the alembic still, which is a design that is very much echoed in modern distillation equipment. Credited as the father of modern chemistry, Al Jabir was responsible for classifying substances by their natures and properties and using distillation to isolate and collect them. He identified a number of acids, hydrochloric, sulphuric and nitric, and left behind a vast archive of writings that explain his process of distillation.

However, in terms of whisky we have to ask, how did all this end up on the Celtic fringes and lead to the fermentation of beer? Initially it was through the works of Robert of Chester, who in 1144 translated the texts of Jabir and others into Latin, but all this brings us no closer to discovering who it was that first started to distil beer in order to produce

Modern stills at the Pearse Lyons Distillery. *Pearse Lyons*

spirits. Although it is widely accepted that distilling beer is a Celtic invention, the first mention of the process in print comes in Chaucer's *The Canterbury Tales* written between 1377 and 1400. In 'The Canon's Yeoman's Tale', the narrator reveals the slippery, elvish art practised by his master. The talk is of alembics and the complex list of ingredients includes oil of tartre, alum gas, berme, wart and beer distilled.

It would be another century before a concrete reference was finally made to grain being distilled in Scotland. Here we are again with Friar John Cor, perhaps the most widely known and yet invisible Scotsman in whisky history. We know next to nothing about John Cor. When was he born, we know not. Where he was domiciled and when he died is also lost to history. In fact, all we know of this man is a single reference made in 1494's exchequer rolls of King James IV of Scotland. That reference mentions eight bolls of malt being delivered to the friar in order to make aqua vitae, the water of life, and as we established earlier, this was enough raw material to produce 790 gallons. This reference is widely agreed to be the first written reference to actual Scotch whisky.

James IV was notorious for campaigning with his men; he would throw himself into sieges and battles and would celebrate with the ranks in victory. Historical records tell us that he was particularly interested in alchemy and was not a man to disallow himself the finest things in life, so it is no surprise that he had a liking for the aqua vitae, which it is believed he discovered while campaigning against the Lords of the Isles on the west coast of Scotland. However, it may be that the Irish actually beat the Scots in producing aqua vitae, but rather than the water of life it would prove to be aqua mortis for one notable imbiber of the spirit. Richard Mac Raghnaill of the Irish Isles was the head chieftain of the Muintir Eolais and during the Christmas celebrations of 1405 he got extremely drunk and didn't wake up the following morning. The coroner reported his cause of death to be 'consuming a surfeit of aqua vitae'.

Aqua vitae, meaning the water of life, was a name given to any distilled spirit but since there is documented evidence that the Muintir

Eolais created their drinks from a grain mash then we know that what he was drinking was whiskey or rather a crude form of the drink we know today. The drink would not have been matured and it is most likely that it would have been a high-proof whiskey similar to the moonshine that became prevalent in the modern age.

Whisky historians tend to debate whether true whisky production actually started with the Scots or the Irish, but given the evidence available it does seem that Ireland has the strongest claim to being the birthplace of the modern liquid that is consumed across the world today, although Scotland certainly wasn't far behind. The authorised version of whisky history states that whisky was at first a medicine and that it took centuries before it became a recreational drink, but whatever the truth, wherever the drink originated, it is a fact that whenever modern drinkers raise the glass and swallow that liquid sunshine, they are tasting a rich and varied history, the bloom of the land, a taste of waters that have passed through mountains and flowed over stone moulded by the passing of time. It is not at all fanciful to say that we are tasting history in a glass.

> *Usquebaugh ... preferred before our own aqua vitae because of the mingling of raysons, fennell seede and other things, mitigating the heat then making the taste pleasant* [sic].
>
> Fynes Moryson, 1617

The above quote comes from gentleman traveller Fynes Moryson, an Englishman who spent much of the 1590s travelling first across Europe and then through the Mediterranean islands. He later wrote about his discoveries in his multi-volume itinerary, which is of extreme value to historians as it so clearly depicts social conditions of the period. He had been educated at Peterhouse, Cambridge, and from 1591 to 1595 he had travelled around continental Europe with the aim to record local customs, institutions and economics. It was in one of his volumes that he wrote of the new spirit he had discovered in Ireland that was very much an early form of the whiskey we enjoy to this day. Then, as now, it is believed the Irish triple distilled their whiskey, which resulted in a smooth and flavourful drink.

> *A man of the Hebrides ... as soon as he appears in the morning, swallows a glass of whisky, yet they are not a drunken race ... but no man is so abstemious as to refuse the morning dram, which they call a skalk. The word comes from the Gaelic which means a blow to the head.*
>
> Doctor Samuel Johnson, 1773,
> after sampling an early form of Scotch whisky

Whisky distillation had by this period become commonplace in both Ireland and Scotland and by the middle of the eighteenth century the process had shifted from small-scale production to commercial distillation. There followed a tsunami of legislation, as governments tried to control excessive consumption while, more importantly at least from their point of view, maximising the revenue gained.

In Scotland the first tax on whisky occurred in 1644, which resulted in many distilleries going underground to avoid the beady eye of the taxman, and a similar situation existed in Ireland. In both countries distillers would ship their whisky out during the dead of night, operating with only the light of the moon to guide them, which is where the term moonshiners originated. The tax imposed was 2s 8d for each pint of whisky produced by a distillery, and increases occurred half a dozen times during the following sixty years. This meant that the Lowland distillers, those within easy reach of excise men, were often operating at a loss and those that didn't go out of business were forced to buy inferior grain in order to cut costs. The Highland distillers, who operated illicitly, bought their cereal from local farmers, which was often of better quality and had easy access to water from the natural springs flowing through the glens surrounding their operations. Following the Act of Union in 1707, the parliaments of Scotland and England were as one, which meant that the same taxes on distillation applied to both countries. This meant that duties on many products north of the border increased sevenfold. In 1715 the English malt tax came into effect and this was increased significantly in 1725.

One of the most famous tax collectors of this period was Scotland's beloved bard Robbie Burns himself, who in 1785 wrote the poem 'Scotch Drink', an ode to whisky and the nature of happiness.

Gie him strong drink until he wink,
That's sinking in despair;
An' liquor guid to fire his bluid,
That's prest wi' grief and care:
There let him bowse, an' deep carouse,
Wi' bumpers flowing o'er,
Till he forgets his loves or debts,
An' minds his griefs no more.

Let other poets raise a fracas
Bout vines, an' wines, an' drucken Bacchus,
An' crabbit names an' stories wrack us,
An' grate our lug:
I sing the juice Scotch bear can mak us,
In glass or jug.

O thou, my Muse! guid auld Scotch drink!
Whether thro' wimplin' worms thou jink,
Or, richly brown, ream owre the brink,
In glorious faem,
Inspire me, till I lisp an' wink,
To sing thy name!

Let husky wheat the haughs adorn,
An' aits set up their awnie horn,
An' pease and beans, at e'en or morn,
Perfume the plain:

Leeze me on thee, John Barleycorn,
Thou king o' grain!

On thee aft Scotland chows her cood,
In souple scones, the wale o' food!
Or tumbling in the boiling flood
Wi' kail an' beef;
But when thou pours thy strong heart's blood
There thou shines chief.

Food fills the wame, an' keeps us livin';
Tho life's a gift no worth receivin',
When heavy-dragg'd wi' pine an' grievin';
But oil'd by thee,
The wheels o' life gae down-hill, scrievin',
Wi' rattlin' glee.

Thou clears the head o' doited Lear,
Thou cheers the heart o' drooping Care;
Thou strings the nerves o' Labour sair,
At's weary toil;
Thou ev'n brightens dark Despair
Wi' gloomy smile.

Aft, clad in massy siller weed,
Wi' gentles thou erects thy head;
Yet humbly kind in time o' need,
The poor man's wine:
His wee drap parritch, or his bread,
Thou kitchens fine.

Thou art the life o' public haunts;
But thee, what were our fairs and rants?
Ev'n godly meetings o' the saunts,

By thee inspir'd,
When, gaping, they besiege the tents,
Are doubly fir'd.

That merry night we get the corn in,
O sweetly, then, thou reams the horn in!
Or reekin' on a New-Year mornin'
In cog or bicker,
An' just a wee drap sp'ritual burn in,
An' gusty sucker!

When Vulcan gies his bellows breath,
An' ploughmen gather wi' their graith,
O rare! to see thee fizz an' freath
I' th' lugget caup!
Then Burnewin comes on like death
At every chaup.

Nae mercy, then, for airn or steel:
The brawnie, bainie, ploughman chiel,
Brings hard owrehip, wi' sturdy wheel,
The strong forehammer,
Till block an' studdie ring an' reel,
Wi' dinsome clamour.

When skirlin' weanies see the light,
Thou maks the gossips clatter bright,
How fumblin' cuifs their dearies slight;
Wae worth the name!
Nae howdie gets a social night,
Or plack frae them.

When neebors anger at a plea,
An' just as wud as wud can be,

How easy can the barley-brie
Cement the quarrel!
It's aye the cheapest lawyer's fee,
To taste the barrel.

Alake! that e'er my Muse has reason,
To wyte her countrymen wi' treason!
But monie daily weet their weason
Wi' liquors nice,
An' hardly, in a winter season,
E'er spier her price.

Wae worth that brandy, burnin' trash!
Fell source o' monie a pain an' brash!
Twins monie a poor, doylt, drucken hash
O' half his days;
An' sends, beside, auld Scotland's cash
To her warst faes.

Ye Scots, wha wish auld Scotland well!
Ye chief, to you my tale I tell,
Poor, plackless devils like mysel!
It sets you ill,
Wi' bitter, dearthfu' wines to mell,
Or foreign gill.

May gravels round his blather wrench,
An' gouts torment him, inch by inch,
Wha twists his gruntle wi' a glunch
O' sour disdain,
Out owre a glass o' whisky-punch
Wi' honest men!

O Whisky! soul o' plays and pranks!

Accept a Bardie's gratefu' thanks!
When wanting thee, what tuneless cranks
Are my poor verses!
Thou comes – they rattle i' their ranks,
At ither's arses!

Thee, Ferintosh! O sadly lost!
Scotland lament frae coast to coast!
Now colic grips, an' barkin' hoast
May kill us a';
For loyal Forbes' charter'd boast
Is ta'en awa!

Thae curst horse-leeches o' the' Excise,
Wha mak the whisky stells their prize!
Haud up thy han', Deil! ance, twice, thrice!
There, seize the blinkers!
An' bake them up in brunstane pies
For poor damn'd drinkers.

Fortune! if thou'll but gie me still
Hale breeks, a scone, an' whisky gill,
An' rowth o' rhyme to rave at will,
Tak a' the rest,
An' deal't about as thy blind skill
Directs thee best.

During the 1820s, in Scotland alone more than 14,000 illicit stills were confiscated by the taxman, and it was estimated that more than half the whisky enjoyed across the country was being done so without the taxman taking a cut. This illicit spirit was being produced in far-flung areas of the Highlands, where it was virtually impossible for the tax collectors to patrol. And yet as arduous as it was, it proved extremely lucrative for the producers to ship out during the midnight hours.

The situation in Ireland was similar and in 1779 there were 1,228 distilleries registered but that same year, when the law taxing stills on their capacity was introduced, 982 of them simply disappeared. The remaining legal distilleries, all in easy to find places, were producing whiskey, as in Scotland, that was inferior to that coming from the illegal stills situated in hard to identify rural areas, which drove up demand for the illicit spirit. Moonshine became the drink of choice, while the legal version suffered in both reputation and sales. This subject will be covered fully in the next chapter.

> *Never delay kissing a pretty girl or opening a bottle of whisky.*
>
> Ernest Hemingway

So, once again we have to ask ourselves, was it the Irish or the Scottish that we have to thank for the modern drink? This will always be argued over and the true origins of whisky remain shrouded in mystery, and likely always will. Legend tells us that distilling was brought to Ireland by St Patrick, but he had grown up in Roman Britain during the fifth century and there is no record of the distillation of alcohol anywhere in the British Isles during this period. Roman writers produced scores of works that detailed the production and appreciation of fine wines, yet there are no records of distilled spirits in any of the texts. So, it remains unclear how St Patrick would have obtained the knowledge of distillation, much less taught this to the Irish. The fact is that if distillation had been practised in the Celtic world, then the Romans would have come across it during their conquests. There is, however, no record of alcohol distillation anywhere in the Celtic world in antiquity. It is far more likely that it was a band of Christian monks who originally brought the art of distillation to Ireland, something they would have picked up on their travels. However, St Patrick will always be associated with Irish whiskey and the Irish Whiskey Association claim that more than 5 million bottles of it are consumed during the week of the annual St Patrick's Day celebrations.

The Irish are proud that their evidence of distillation predates that of Scotland, and there are some claims that the Irish practised distillation as far back as the sixth century but this is unsubstantiated and, as previously noted, the wealth of evidence does suggest that it is likely whisky distillation started in Ireland and that the knowledge was then brought to Scotland by those travelling monks crossing the 12 miles of water between the North Antrim coast and the Mull of Kintyre.

However, even this is far from certain since there are as many differences as there are similarities between Irish and Scottish forms of the drink. If Scotch did develop from the Irish form of the drink, then the flavours would likely be far more similar. If Irish monks did introduce distillation to the Scots, then this does not explain why the Irish distil their spirit three times, while the Scots as a general rule only do it twice. Likewise, the Scots dry their malted barley using peat fires, which imparts a smoky flavour to the drink, while the Irish use a smokeless kiln to dry their barley. The Irish used a mixture of both malted and unmalted barley, while the Scots only used malted barley, or at least that was the case until the industrial revolution when Lowland distillers started to introduce unmalted barley into their whisky production. If Irish monks had brough distillation to Scotland then surely they would have taught the same process used in Ireland, so we remain far from certain that this was the case.

The exact origins of whisky distillation in both Ireland and Scotland are obscure, and short of some monumental historic discovery this will forever remain so.

Chapter 2

By the Light of the Silvery Moon

For many of us the word 'moonshiner' will sum up the clichéd image of an American hillbilly, likely named Jed or Cleetus, three teeth in mouth, banjo over shoulder and scattergun in hand. While we will meet such colourful characters later in this book, we have already established that the origins of moonshine go back much further in both time and place and that it originated oceans away in Scotland and Ireland. We have noted that the first tax was placed on whisky in Scotland in 1644, which meant that many distillers decided to flout what was largely seen as an unjust levy, and simply took their production underground.

Farmers in remote areas had long been using small stills that held around 40 gallons to distil whisky and were bringing in a good income from doing so. They were loath to hand over any of their gains to the excisemen. Invariably these farmhouse stills were situated in the Highlands, whose crags and glens were difficult to reach by tax collectors, and so the farmers came up with numerous ingenious ways to transport their whisky and get it into the hands of their customers. Mostly the illicit stills were concentrated in Glenlivet, Kintyre and Islay, which were all remote inhospitable areas that were difficult to patrol.

Scots openly defied the taxes, and many of those in Scotland employed to collect them ignored much of the illicit activity, often turning a blind eye. Scotch whisky in particular is steeped in rebellion, riots, smuggling and quite often violence, and all in the name of avoiding taxes and creating a good-quality beverage. Many otherwise law-abiding Scots were more than willing to commit crimes in order to produce whisky without government interference and industrialised ideas that many believed would change the character of their product. The drink was ingrained in the Scottish way of life, used in weddings, funerals and christenings. It was often used as medicine to aid the recovery of the sick and wounded, and the high taxes that the government sought to impose threatened not only whisky but Scotland itself. It challenged a long-cherished way of life.

Prior to the Act of Union in 1707, English customs men were not at liberty to cross into Scotland and enforce taxes and so Scotsmen themselves had to be tasked with the role. The threat of violence, and perhaps a bladder or two of fine whisky, would ensure that they carried out their job in a half-hearted fashion. However, with the Act of Union and the resulting unification between Scotland and England, the government decided to strike at the illicit whisky market with a vengeance. The government was now at liberty to send English revenue men across the border to collect taxes and come down hard upon illicit stills. The Scottish, though, would not be manhandled and were not going to give in without a fight. Often when excisemen discovered an illegal still they would be attacked by mobs and driven from the areas they were supposed to be policing. Smuggling in Scotland was almost a universal pastime. The majority of people saw taxation as an imposition upon ancient liberties and they were not going to accept it without a fight. More power to them, I say.

A new phrase, the bladder men, came into use as smugglers would fill a sheep's bladder with whisky, hide it beneath their britches and deliver it to customers who had by now realised that Highland whisky was far superior to that being produced by the legal distilleries in the Lowlands. This was largely due to the legal distilleries having to cut production costs in order to survive, and one way to do this was to purchase inferior grains and malt with which to produce their goods. This, of course, resulted in a harsher dram than the cream coming from the Highlands, which further drove up demand for the illicit whisky. Whisky smuggling became even more lucrative and the manner in which the whisky was transported became ever more ingenious. Women would be fitted with belly canteens that held a couple of gallons of whisky and gave them the appearance of being heavily pregnant, while bottles of whisky were taken to markets hidden within dead geese or turkeys.

In 1725 the government changed tack and rather than trying, and largely failing, to tax whisky by the volume made by a particular distiller, it decided to go the root of the problem and place a tax on the malt used to produce the drink. This was set at 3d for each bushel of malt, which meant that distillers could no longer avoid the tax without roping in someone with a malting facility. This is pretty much exactly what happened.

The Scottish were now united in their anger against the government across the border, and the new taxes unified both those in the whisky and agricultural industries into a common goal – stick it to the English! Many farmers, millers, and roasters revolted against the newly levied taxes. Some took to hiding their products or falsifying their books, and a great many flatly refused to allow tax inspectors to examine their wares by concealing them or often physically driving collectors away with violence. Soon resentment grew between the Lowland and Highland distillers because the efforts of the tax collectors was not carried out equally between both regions. This only added to the already strained relations between the two regions. The Highlanders considered the anglicisation of the Lowlands as treachery against Scottish heroes such as William Wallace and Robert the Bruce, who had fought against English rule in the thirteen and fourteenth centuries. In turn, the Lowlanders often considered the Highlanders to be uncivilised barbarians, which was often a view shared by the English, especially the excise men who had suffered at their hands.

> *Those who drink whisky with the owls at night, cannot soar with the eagles the next day.*
>
> Brian D. Ratty

The Lowlands were closer to the English border and that made it easier for the tax collectors to target their activities, and thus the Lowland distillers made less profit per bottle than those situated across the Highland boundary fault. The Lowlands, while far more agriculturally suited to malt growing, didn't have the defences of the varied terrain, including huge mountain ranges to deter the tax collectors. Added to their woes was the fact that Highland whiskies were usually of better quality, given that the high pressure in the mountains allowed the whisky to mature quicker and the topographical variety of the region allowed for a far more varied range of whisky expressions. This meant that Lowland distillers, forever operating under the eye of the excisemen, had to lower production costs even further. They resorted to cutting the amount of malt they used and making up the difference with less-suitable grains, invariably the cheapest they could find, to distil their whiskies. This resulted in the better-quality Highland whiskies fetching a higher price and being in far greater demand than those produced in the Lowlands. This would plague the distillers for many years as thirsty drinkers eagerly sought out Highland whiskies while largely ignoring those produced in the Lowland region.

The growth of both legal and illicit distilling would continue until disaster struck in 1757 when a massive crop failure led to the government banning the sale of distilled spirits for three years. However, the use of a private still was not prohibited, which resulted in even more illicit whisky making. The law said that such whisky had to be for private consumption, but of course illicit distillers were used to operating outside the law and so this made little difference to their activities, other than taking away the legal competition. As demand grew the smuggling was taking place on a massive scale. More and more women were employed to wear the large whisky-filled containers beneath their clothing, making them appear pregnant, in order to get the newly made whisky to the customer. The smugglers built special coffins to carry whisky, often directly past officials who would bow their heads respectfully as the cortège went by.

The antagonism between the Scots and the excise men would often turn to violence, and confiscating someone's whisky could, and often did, result in bodily harm. There are documented cases of excisemen being kidnapped to keep them from testifying in court against illicit distillers and these men were not always released unharmed once the court case was over. In 1739, Thomas Carswell, an exciseman, was murdered by a smuggling gang while escorting them to prison, and ten years later, in 1749, another excise officer, William May, boarded a suspected smuggling ship and simply vanished without trace. It is believed he was struck over the head and tossed into the sea, his body never found. It was perhaps because of these dangers that many excisemen chose to accept bribes in order to turn a blind eye to the manufacture of illicit whisky. It was a similar situation across the sea in Ireland and Aeneas Coffey, an exciseman who would later invent the Coffey still, barely survived after being stabbed by a smuggler he was attempting to bring to justice.

> *With more insight into the English character I poured out a stiff whisky and soda and placed it in front of the gloomy inspector.*
>
> Agatha Christie, *Lord Edgware Dies*

Perhaps the most famous example of violence as a result of whisky smuggling were the Porteous riots in Edinburgh in 1736. They began after three men, Andrew Wilson, George Robertson and William Hall, were tried and condemned to death for the crime of whisky smuggling. Hall would later have his sentence commuted to exile after he turned king's evidence against his fellow conspirators, while the other two men were not so lucky and awaited their fate in Edinburgh's notorious Tolbooth Prison. Several days before the planned execution the two men were taken to the Tolbooth Church for a pre-execution ceremony, which was customary for condemned men, and during the service the sprightly Robertson took his chance and managed to push his guards aside. While Wilson grappled with the guards, he made a daring dash for freedom. He was successful and managed to flee the church and vanish into the city, never to be captured.

The sentence on Andrew Wilson was carried out in public on 14 April. However, most in the crowd felt the execution was unjust and when a sailor cut down the body of Wilson, Captain Porteous of the City Guard fired at him but missed and killed one of the crowd. This further enraged those gathered and a full-scale riot broke out. The captain, together with his guardsmen and the hangman, had to seek refuge in the city guardhouse on the Royal Mile. The situation worsened and the rioting crowds tried to force their way into the guardhouse. The terrified captain ordered his guards to fire, initially over the heads of the rioters but when this only served to stoke their anger, he ordered his men to shoot directly into the crowd and a further five people were killed. Though there is some confusion over the actual events of the day with many accounts differing, the official records state that in total nine people were killed and more than twenty were left wounded or maimed by the city guards.

Captain Porteous faced condemnation from not only the people of Edinburgh but the country at large, and the city magistrate had no option but to order his arrest. He was tried at Edinburgh's High Court on 5 July and, although he admitted to causing the deaths of six people and the maiming of others, he pleaded self-defence. He claimed that he had feared for both his own life and that of his men. The crowd had grown into an unmanageable number and had been threatening to break into the guardhouse, the gates were holding for the moment but with each surge of the crowd they were shaking in their foundations.

Leading up to the trial the authorities had been alarmed by the depth of feeling against Porteous. It did not help that the captain had already been found guilty by the Scottish press, who were uniformly appalled by his action of ordering his men to fire into the gathered crowds, so it would seem that the only verdict available to the jury was that of guilty, even if this was questionable given the fraught circumstances at the time.

The trial lasted five days and as expected resulted in the inevitable guilty verdict. Porteous was sentenced to death by hanging, with the date for the execution set for 8 September 1736. When news of the verdict spread to the angry crowd outside the court it served to quiet them and most felt that the sentence was the proper one. Porteous was then taken to Tolbooth Prison to await sentence and imprisoned in the same cell that had housed Wilson and Robertson.

However, the events had horrified the government in England and Prime Minister Sir Robert Walpole, and Queen Caroline, acting as regent while King George II was out of the country, granted a reprieve of six weeks as a preliminary to the granting of a full pardon. This only served to anger the people of Scotland, who felt that this displayed indifference on the part of the government to the killing of Scottish civilians. Public feeling would soon reach boiling point and would result in a well-planned plot by a group of Edinburgh men to take matters into their own hands and deal out justice on the disgraced captain.

The general feeling among the authorities was that a mob would attempt to storm Tolbooth, but the politicians did not think it likely and when another captain of the City Guard, Captain Lind, took his fears of a mass protest to the Lord Provost he was told that they were groundless. However, the precaution was grudgingly taken to place extra guards at the prison guardhouse. Word soon got to the authorities, leaked by members of those who were planning to get to Porteous, that a group of men were plotting to lynch Porteous and that they intended to act on 8 September, the date originally set for the execution. However, they outfoxed the authorities by acting a day earlier.

On 7 September at around 9.30 p.m., a mob, said to be in excess of 4,000 men, arrived at Tolbooth Prison and quickly overpowered the guards. They took the terrified Porteous from his cell and practically dragged him kicking and screaming to the Grassmarket, where he was hanged from a dyer's pole using a rope taken from a local draper's shop. After a short while the captain was taken down, stripped naked and then, after having his nightshirt tied around his head, he was hauled up again. This was repeated several times and each time he was brought down the screaming captain would be punched, kicked and even burned by torches before being hauled up yet again. The fury of the mob was absolute and no mercy was shown. Eventually, when it was ascertained that Porteous was dead he was left hanging and the crowd quickly dispersed. For around two hours a mob of around 4,000 had been in charge of the city and had been able to act with impunity.

The events horrified the government in London and it ordered a full inquiry to bring the men responsible for the lynching to justice. However, despite a large reward being offered for information as to who was involved in the lynching, no one was ever arrested.

Today the remains of Captain John Porteous rest in a grave bearing the inscription 'John Porteous, a Captain of the City Guard of Edinburgh, murdered September 7th, 1736'. The actual spot where he was lynched is marked by a memorial plaque, while the location where Tolbooth Prison once stood is commemorated by a group of paving stones arranged in the shape of a heart, known as the Heart of Midlothian. Even today people will spit on the stones; a tradition originally intended to display contempt for the hated Tolbooth.

One legal distillery that flourished despite the competition from the illicit distilleries was Ferintosh, which was situated on the

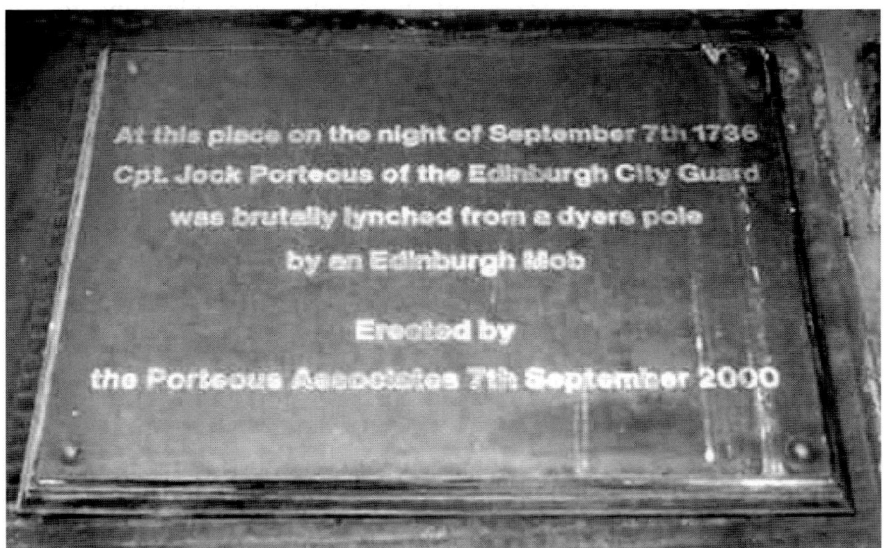

Plaque commemorating the events at Grassmarket. *Author's collection*

Black Isle estate close to Dingwall. It was established in the late seventeenth century and was actually Scotland's first legal distillery. Its dominance during the period was due to the fact that Ferintosh paid absolutely no tax at all, which went back to the death of Queen Elizabeth in 1603.

The Queen, of course, had no children, no heir to take over her rule and so the throne went to her closest relative, King James VI of Scotland. He took the throne and became James I of England, with the Stuart dynasty continuing to rule in Scotland. However, a constitutional crisis in 1688 led to the ousting of the Catholic James II (grandson of James I) in favour of William of Orange. James and his successors continued to lay claim to the thrones of England, Scotland and Ireland and there were periodic uprisings, the most famous of these being the Jacobite Uprising of 1745. Earlier, in 1689, Duncan Forbes, the third Laird of Culloden and the owner of Ferintosh distillery, had declared support for William of Orange, which resulted in the building being attacked and burned down by Jacobite supporters.

Forbes petitioned the Scottish Parliament for compensation and was awarded a generous pay-out. He initially petitioned for £54,000, but the award given was much more generous and would give his distillery a significant advantage over competitors. Parliament awarded him and his family the privilege of distilling grain from their own estate in perpetuity free of duty for only an annual payment of 400 Scottish merks, around £43 today. Ferintosh distillery was written into history and was not slow in making good on this valuable concession, which saved Forbes an absolute fortune and allowed his distillery to dominate over all others for many years to come. Indeed, by the end of the 1690s more than two thirds of all legal whisky sold in Scotland came from the Ferintosh distilleries.

Ferintosh would prosper and his successors continued to do so, and by the 1760s three more distilleries had been erected. They had also acquired more farmland upon which to grow barley and by the early 1770s were producing around 400,000 litres of whisky each year, making substantial profits thanks to their tax concession. The legal status of the distillery allowed them to reach lucrative markets such as those in London and Ferintosh holds the distinction of being the first whisky ordered by name. It would not last though, and in 1784, under pressure from other licensed distillers who rightly resented the concession given to the Forbes family, Parliament moved to withdraw the special status. Soon afterwards the distillery closed forever.

Thee, Ferintosh! O sadly lost!
Scotland lament frae coast to coast!
Now colic grips, an' barkin hoast
May kill us a';
For loyal Forbes' charter'd boast
Is ta'en awa?

Robert Burns, 1785

Ferintosh, though, had been the exception to the rule and most legal distillers, in other words all those that were taxed, found it next to impossible to survive without the income from illicit activities. Even those in the Lowlands, where the risk of discovery was far greater, took to distilling by the light of the silvery moon. Landlords

Vintage whisky label. *Author's collection*

who rented their properties to distillers would often lend a hand in the clandestine activity if it meant their rent would continue to be paid, and those who did not get involved turned a blind eye to the illegal operations being carried out on their properties. Many of these stills would be run by the men while their wives, known as goodwives, would take the whisky to market hidden beneath their clothing.

Whisky distilling became an obsession to the Scottish and the act of defying the English tax system was seen largely as an act of patriotism. A system was soon in force where towns would signal to the next town when the tax inspectors were about. This proved pretty efficient and lookouts would spend their days at strategic points watching for sheets to be hung in pre-determined locations and then sending messengers to warn the next town and the one after that and the one after that … As soon as one still was discovered and dismantled, three more would spring up to take its place and the tax collectors found they had a next to impossible job to carry out; at times it seemed it was them against the entire country. In time many of the Lowland distillers, having made sufficient money from illicit distilling, moved further afield with their operations, up into the Highlands where relative safety was offered to them.

This meant that the English tax inspectors had no option but to face the dangers and travel further into the Highlands, and soon those distilleries that had previously acted without interference from the authorities found that they could no longer continue to do so. Many of the Highland distilleries were so large and conspicuous that they had no option but to become licensed, which meant that they would endure regular revenue checks by government officials. The smaller operations could move their distilleries further into the mountains whenever there was a risk of discovery but such was not the case for the larger distilleries. Soon the tax revenue collected by the government increased to such a rate that the funds were there to employ more and more excise officials to ferret out the illicit operations hidden away in the Highlands. The authorities sent scores of new excisemen out in the search for illicit distillers.

For many years the battle between the tax collectors and the illicit distillers raged and during the late 1700s the government responded by introducing a tax of £30 for each gallon of whisky produced, and it would only go up from there; at one point the tax stood at £160 per gallon. The government believed that the heavier taxation would force the legal distilleries to confront the illicit operators and bring the industry under control. However, this was not the case and the ever-increasing levies simply meant that legal distilleries tried to further cut production costs, which resulted in the quality of their whisky worsening even further, while the demand for the hand-crafted illegal whisky became ever greater.

In 1814, in a last-ditch attempt to confront illicit distilling, the government made a move to ban all stills with less than a 500-gallon capacity. The logic seemed to be that larger stills were harder to conceal from the eyes of the excisemen, but all this ludicrous piece of legislation served to achieve was to increase Scottish defiance to the unfair taxation being imposed by what many saw as a foreign government that had no business interfering with them.

The government finally realised that its efforts to stamp out illicit distillation by force was next to impossible; that it was fighting a battle they could not win. The illicit distillers were highly organised and worked mostly with the full support of their communities. Often clergymen would even help out by allowing barrels of whisky to be hidden beneath their pulpits to be later moved out within specially adapted coffins. The people joined together to avoid detection and gangs of smugglers would be made up of men, women and children, and more often than not would easily evade the excisemen.

Nor did huge fines imposed by the courts stop the illegal smuggling of whisky. By 1812 the government had raised the fine for illicit distilling and smuggling to £20, a fortune during the period, but the high profits made from the whisky was an even greater fortune so the risks, which had always been minimal in any case, were worthwhile. Illicit stills were often run as a cooperative with the community, so all shared in the profits and would band together to pay any fines the whisky produced.

> *Laws against smuggling are generally disliked, and Scots will cheerfully break the law. When I was young, everyone I met from my father downwards, even our clergy, either made, bought, sold, or drunk whisky that had been smuggled.*
>
> Dr John Mackenzie, *Highland Memories*

In 1823, the government, which had long been haemorrhaging money in its war against illicit whisky, decided to do away with the taxation system and introduce a modest licensing fee in its place. It legislated that only stills of 40 gallons and above would have to register for a licence, which had the effect that all of the decades of heavy taxation could not achieve. Gradually the illicit distillers decided it was too much trouble to remain hidden and they emerged and embraced their newfound legal status. One such illegal distiller who took out a licence was George Smith, a farmer who had been illicitly producing a quality whisky from his small distillery for a great many years. The whisky he produced was Glenlivet and today the distillery that started from that small operation produces more than 10 million litres a year.

Scotch whisky has come a long way since the days of unfair taxation and illicit production, and today this fine malted beverage enjoys access to a global market. There are some great whiskies being produced around the world, but Scotch continues to dominate and is the most recognisable type of whisky in the world. Scotch is these days recognised and protected by legislature as the Scotch Whisky Regulations 2009 determines what Scotch is, how is must be labelled, and even sets parameters for its advertising.

The basic rules for Scotch are:

1. It must be composed of only water, malted barley and other grains, and the distiller may only add yeast and some caramel colourings if needed.
2. It should be distilled to a proof of no less than 94.8%.
3. It must be aged in oak casks with a capacity of no more than 700 litres, for a period of no less than three years. This ageing must be carried out within Scotland. There was a time when it could be partly aged elsewhere, but not any longer. Today all Scotch must be aged fully in Scotland itself.
4. It must be determined to have the colour, aroma and taste of the grains used in the distillation still intact.

Chapter 3

Trouble on the Emerald Isle

As stated earlier, it is believed, though not ascertained, that the Irish were likely the first people to discover whiskey distillation but whatever the truth it remains a fact that, alongside the Scottish, the Irish were there during the infancy of whiskey production.

However, the whiskey produced back then was not the delicious amber drink we know today, but would have been clear, unaged and rather harsh on the palate. Drinkers would flavour the drink with herbs, spices, fruits and berries, making it closer to gin than whiskey. It is also firmly established fact that in 1170 when King Henry invaded Ireland the troops came upon the aqua vitae that the Irish were producing and enjoying. This tells us that whiskey production was widespread enough by this period to be recognised as a part of regular society. The Irish, as did the Scots, found themselves under English rule and as such were subject to similar taxation. And like the Scots, the Irish were not keen on having what was seen as a God-given right made the subject of taxation. However, unlike the Scots the Irish had the advantage of having an expanse of water between them and their mainland rulers.

In 1556, the English Parliament passed an act that restricted distillation without a licence in Ireland, but most ignored it and distillation continued apace. In fact, whiskey now experienced great growth, spreading to all corners of the country. Unlike expensive wines, whiskey was not solely the drink of the upper classes but was shared by all. When the government threatened martial punishment for anyone who violated the act and continued to distil without a licence, the Irish also decided to ignore this. It was difficult to see how the English government, who were tied up with so many conflicts, could make good on their threats.

Indeed, it would not be until 1608, decades after the initial act had become law, that the first licence to distil was issued. This was granted to Sir Thomas Phillips, and his Old Bushmills Distillery became not only the first ever distillery to be licensed in the country, but actually the first in the world. Today, the distillery makes a big noise about this distinction in its advertising. However, as with everything related to whiskey, there is some argument to the veracity of its claim to be the world's oldest licensed distillery, since while it is true that the first ever licence to distil was issued to Sir Thomas in 1608, the Old Bushmills Distillery was not founded until 1784, and has only been in continuous production since 1885, having been rebuilt after a fire destroyed the original distillery. This technically makes the Kilbeggan Distillery, founded in 1757, the oldest licensed distillery.

Vintage Bushmills advert. *Author's collection*

The Bushmills argument, though, will never be settled, and the company continues to sell its whiskey as coming from the world's oldest licensed distillery. However, what is certain though is that Irish whiskey boomed for a period, but then in 1648, with the outbreak of the English Civil War, it came up against the first of many setbacks that would eventually all but wipe out whiskey production in on the Emerald Isle. Indeed, it was not until the mid-1980s that Irish whiskey saw the first stirrings of the renaissance that would see the drink reinvent itself and bring it to the healthy position it enjoys today in the world market.

The Civil War, and especially its aftermath, though damaging to the industry would not bring the ruination on Irish whiskey that later conflicts would, but it was nevertheless a blow. When Oliver Cromwell died in 1658, his son stepped in as Lord Protector, though he had little knowledge of either politics or military command. He soon ceded control to Parliament, which created a power vacuum and caused chaos across Britain. The answer to this was to restore Charles II to the throne, and given that the war had been expensive, the King, together with Parliament, placed a tax on whisky production across Britain and Ireland of 4d on each gallon produced.

The taxation meant that illicit distilling surged, since the legal distillers found that they would either have to take the loss themselves or pass it on to the customer, and neither option was ideal. Given that the Crown's influence in Ireland didn't stretch much beyond Dublin, this left the rest of the country to do as they pleased regarding whiskey production without too much interference from English tax collectors. As in Scotland, illicit whiskey was also favoured over the legally distilled variety, which was termed, rather sarcastically, parliamentarian whiskey. Smuggling, as it had in Scotland, became the occupation of freedom, a way to hit back at what was seen as a foreign power trying to steal from the indigenous people.

The Irish experience regarding illicit whiskey, although similar to the Scottish as detailed in the last chapter, differed because of the Irish Sea that separated the country from mainland Britain. And while

Irish whiskey continues to thrive. *Pearce Lyons*

it is true that excisemen found it difficult to tackle many of the illicit distillers situated in the more remote parts of the Scottish Highlands, it was next to impossible to subdue the illicit distillers across the expanse of sea. The answer to this was to recruit excisemen from within Ireland, but these men usually carried out their job in a half-handed fashion rather than risk attack or even murder from the mobs of men who would arm themselves in order to protect their stills and the whiskey they produced. Communities would band together, forming lookouts to warn of the approaching excisemen as well as often arming themselves and engaging the terrified tax collectors.

The manufacture of whiskey was largely seen as a peasant industry in Ireland, and indeed until 1760 it was legal for each household in Ireland to distil enough whiskey for its own use. This was important to agriculture since the production of the drink demanded a steady supply of grain, and the sale of the whiskey produced helped tenants to pay their rents, which kept the landlords on the side of distillers. When laws were brought in it meant that overnight private distillation became illegal, but in the eyes of the majority of the people it was the laws that were wrong, not the distillation.

In 1779 the government introduced a new tax that was set on the capacity of the stills used by the legal distillers. This tax was based on the production capacity of the still and worked on the assumption that a still was running each and every day of the year. This meant that the tax was placed on the amount of spirit it was possible to produce rather than the actual production of a given distillery. This was a lazy way of streamlining taxation and only resulted in even more illicit distilling. Then, in 1783, the government introduced a law that would fine all members of a community in which an illegal still was found. The hope was that communities would inform on illicit stills, but in reality it only strengthened the ill feeling towards the government. Before the act of 1783 there were 1,200 legal stills registered in Ireland but this quickly dropped to only thirty by 1807 as more and more went out of business, though strangely the amount of whiskey in circulation greatly increased. The licensed stills that did continue were all based in urban areas, which had the result that those in the hinterlands could only be served by illicit distillation.

For the modern historian looking back it is plain to see how absurd the laws against distillation were. The aforementioned law of 1779 meant that licensed distillers could no longer make enough profit to keep them in operation or at least operating legally. Instead of visiting each still to gauge the amount of spirit produced and then applying taxation fairly, the authorities rather lazily estimated the least amount of spirit each still could possibly produce and then charged duty on that estimated amount, whether production actually reached that level or not. However, this meant that by speeding up distillation, a distiller could produce far in excess of the minimum level and thus much spirit could be produced free of duty. When the authorities realised this, they simply increased the estimated minimum levels, which resulted in even more rapid distillation, but many distillers found they could not compete and so went out of business, or rather disappeared from the licensed trade.

Another provision of the 1779 legislation hampered the use of small stills, and stated that all must have a minimum capacity of 500 gallons. Many campaigned against this restriction and a reform in 1809 removed the rule, but it had little effect on illicit distillation since the Board of Excise stubbornly refused to grant licences to small stills. This reluctance can be partly explained by a Treasury minute of 1812 made by the Commissioners of Revenue, which instructed licences to be issued to small stills only in very exceptional circumstances since it was plainly possible under the current system for a them to be used to defraud the Revenue of large sums of money. It was noted that in Scotland many licences had been issued to small stills, but the Scots, by speeding up distillation, had produced quantities of spirits that far exceeded the amount estimated for taxation.

Wee devils, those Scottish!

The war between excisemen and the illicit whiskey producers continued for many years. Indeed in 1810, Aeneas Coffey, who would later invent the Coffey still that is still used in whiskey production to this day, was an exciseman. He had been appointed surveyor of excise for Dublin in 1809 and in 1810 he transferred to County Donegal, where illicit distilling was rife. It was here on 8 November 1810 that he was brutally attacked, bayonetted and then left for dead by an angry group of illicit distillers. He, of course, survived his great injuries, but his attackers were never brought to justice and he received little sympathy from the general populace, since the excisemen were extremely unpopular in areas where agricultural income had long been supplemented by whiskey making. Coffey would rise to become inspector general of excise, but in 1824 he resigned his position and later purchased 800 acres of land in County Kildare. It was here that he patented a new still he had designed that was fitted with hydrometers. He was awarded a patent in Ireland that was to last for fourteen years, but his still was not widely adopted by the Irish, partly because its iron pipework initially left an unpleasant taste to the spirit produced, and no doubt partly because his past role as an exciseman was resented.

Coffey would move to London in 1835 and in the 1840s the Coffey still was adopted by Scottish distillers. The Coffey still has been used worldwide ever since, in vodka as well as whisky production. The stills are particularly energy efficient, and do away with some of the stages of production, which allows for non-stop operation and produces a flavourless spirit, which is the main ingredient of Scotch whisky blends. The Irish, faithful to their pot stills, made many unsuccessful attempts to challenge the use of the word, 'whiskey' to describe this blended spirit, but eventually the Coffey still would be adopted by some in Irish whiskey production.

In 1823 the English finally realised that they were fighting a losing battle and they introduced a law where it would cost £10 to obtain a licence to distil, as well as a small taxation on the amount of whiskey actually produced rather than the production capacity of a given still. This gradually saw the illicit distillers disappear as more and more became licensed, for while taxation was still resented it was seen as a necessary evil and preferable to constantly hiding away from the attentive eye of the excisemen.

Once more all seemed well for Irish whiskey, but the greatest challenges were yet to come.

Perhaps the earliest mistake the Irish whiskey industry made was in passing on the continuous still invented by Aeneas Coffey in favour of their traditional pot stills. While it is true that the continuous still enabled alcohol to be produced more efficiently and in far greater volume, the Irish felt that it stripped away too many impurities and created a blander whiskey than that produced in the traditional pot stills. This was the reason the Irish producers gave for rejecting the continuous still though, as previously speculated, many were still angered by Coffey's past as an exciseman. The Scottish, however, had no such scruples and they immediately adopted Coffey's invention and found themselves able to produce previously unfathomable amounts of the drink, which allowed them to flood the market and feed an ever-growing global demand for Scotch whisky. A demand that kept on growing.

It was during this period that the Scottish also started blending, then called rectifying, their whiskies, mixing older spirits with younger spirits to create a consistent flavour profile and offer the product to the market at a much cheaper price than the Irish were able. Blending slowly became an art and the demand for these fine spirits surged, leaving the more expensive Irish variety struggling to compete. For much of the nineteenth century, Ireland had dominated the market, with more than 70% of the world's whiskey coming from the Emerald Isle, but by the end of the century the Irish had been overtaken by the Scots, largely because of the invention of the Coffey still that boosted output and lowered costs. It must have rankled that the invention was the work of an Irishman and to add insult to injury a man who in his previous role as an exciseman had once tackled the illicit stills that provided much of the really good whiskey. Eventually though the Irish would be forced to adopt the continuous still, but not before Irish distilling was brought to its knees.

The twentieth century would bring hardship after hardship to the Irish whiskey industry and no sooner had they found an equilibrium within the world market than it was torn asunder by the outbreak of the First World War. At the outbreak of the conflict, Ireland produced more than 10 million gallons of quality whiskey a year, but one of the major markets for Irish whiskey was America and the threat posed by German U-boats made it hazardous if not impossible to export the sprit. The vital international market was cut off and the demand for Irish whiskey plummeted.

No sooner had the war ended than Ireland was once again thrown into turmoil by the war for independence. There had long been a core faction within Ireland who resented British rule, and indeed in 1916 the Easter Uprising had seen a rebellion against the British in Dublin when a group of rebels commandeered strategic buildings in the city and engaged the British in a six-day stand-off. Soldiers, artillery and a gunboat had met the rebellion head on and sixteen rebels were executed by the British and more than 3,000 Irishmen were taken prisoner. During the skirmish more than 400 people were killed, 300 of these being civilians.

> *Small boys bolted in to see these sights and bolted out again with bullets quickening their feet. Small boys do not believe that people will kill them, but small boys were killed.*
>
> James Stephens, 1916

During the war for independence, the Irish economy turned inward, and the resulting full-blown conflict against the British saw many of the major distilleries occupied by either British or Irish forces. When they departed the distilleries were, more often than not, left severely damaged or in several cases completely demolished. The eventual ceasefire in December 1921 saw Ireland become a self-governing dominion, but Northern Ireland had declared itself its own territory and remained under British rule, effectively segmenting itself from the rest of the country. This would cause tensions that to some extent continue to the present day. There was also the fact that trade embargoes had been imposed by the British that prevented Irish exports to the Empire including Canada, South Africa, India, New Zealand and Australia.

> *When life hands you lemons make whiskey sours.*
>
> W.C. Fields

A modern Irish blend with a rich history. *Author's collection*

The next major stumbling block for Irish whiskey was American prohibition, introduced in 1920, and this would virtually decimate the already troubled industry. Prior to the prohibition act Ireland had eighty-eight licensed distilleries and produced around 12 million cases of whiskey a year, but over the first few years of prohibition thousands lost their jobs in the industry as distillery after distillery went bankrupt. The Irish, unlike the Scots, were unable to export their spirits to Canada because their independence meant that they had lost trade rights with British colonies. Much of the Scotch sent to Canada had been illegally re-exported into America, but in light of the embargo Americans, always big consumers of Irish, now found their tastes changing to prefer the Scottish variety. When prohibition ended in 1933, the whiskey industry in Ireland was all but a shadow of its former self and for decades there were fewer than a handful of Irish distilleries.

Irish whiskey had by this time suffered blow after blow. The temperance movement of 1838 led by Father Theobald Mathew, a Capuchin friar, had gained a stronghold across the country and greatly impacted on whiskey consumption. There was also the fact that the great famine of 1845–49 had seen over a million die of starvation and the emigration of another million, which caused the country's population to fall by 25%. And now with all the political upheaval of the new century Irish whiskey was finally brought to its knees. War, famine, temperance and then the fierce competition in the market from the Scottish with the upsurge in the popularity of blended Scotch all seemed set to deliver the final nails to the coffin of Irish whiskey.

However, Irish whiskey would thankfully prove too tough to die, and its eulogy was somewhat premature. Today there are over forty distilleries in Ireland and the spirit has seen something of a resurgence, with new distilleries cropping up to continue the long heritage of whiskey production on the Emerald Isle as the demand for Irish whiskey continues to grow.

The resurgence started in 1966 when the three remaining Irish distillers, Jameson, Powers and Corks, joined forces to form the Irish

Distillers Company. They would later acquire Bushmills, the last distiller in Ulster. However, the company failed to meet the demands of the shareholders and soon a takeover loomed. The French group Pernod Ricard would eventually take over the company and flex its massive marketing might in pushing two of the numerous brands it had acquired, Jamesons and Bushmills. Irish whiskey began its first tentative steps towards regaining a foothold on the world stage.

Welcome back, old friend.

Chapter 4

A Whisky Mixer

Since you are reading this book, it's a fair bet that you are somewhere along your whisky journey. Perhaps you are a seasoned imbiber, one who knows their malts from their grains and are familiar with the nuances of different whisky expressions, or maybe you are taking your first tentative stumbles towards a better understanding of the complexities of the wonderful drink celebrated within these pages. It matters not, for both novices and those well versed in the mysteries of the amber gold know that there are few pleasures better than sipping a dram while listening to or reading many of the stories and legends that are an intrinsic part of the world of whisky – whiskey, *uisge beatha*, *uisce beatha*, *usquebaugh*, but let's not start that again.

This miscellany, together with a second section later in this book, collects together many of the stories, legends and perhaps downright lies that are well documented in the annals of whisky appreciation. Here you will learn of ghosts, murderous cats, dalliances with the devil and so much more.

So, pour a dram, settle back and let's take a journey down roads and byways seldom trodden.

A Toast to the Ghost

The Glenrothes Distillery, situated in Scotland's Speyside, holds the distinction of being the region's most haunted distillery and to this day whenever a new batch of whisky is tasted for the first time a tradition is observed in the form of a toast; a toast to the ghost of an African man named Biawa 'Byeway' Makalanga. So how did a distillery situated in the village of Rothes come to be associated with the ghost of an African man? And why should the restless spirit choose to walk the darkened corridors of Glenrothes, when in life Byeway was associated with the Grant Distillery, which is situated at the northern end of the village?

Well, that's quite a story.

Glenrothes itself is situated across the road from a cemetery where the gravestones are blackened by a fungus that thrives around many of Scotland's distilleries, stubbornly clinging to the old brickwork of the walls and spreading at an alarming rate. The mould is known as Baudoinia and is harmless but looks unsightly, which angers those who live close to distilleries. Indeed, in 2014 residents of the Scottish town of Beith brought a legal case against global drinks giant Chivas Brothers because the mould was colonising their homes.

It is believed that when the airborne spores of the fungus meet with the ethanol vapours, known as the angel's share, which evaporate from whisky distilleries, this causes the mould to grow at a rapid rate. A similar situation occurred in Tennessee when in 2023 the Jack Daniel's company was forced to halt the construction of a new barrelhouse because of complaints from residents that the black mould was covering buildings close to the distillery. The fungus itself grows on distilleries across the world and thrives where fermentation occurs, such as in bakeries and distilleries.

It was first discovered around the Cognac region of France in 1872 by a scientist named Antonin Baudoin, who described it as a plague of soot. It was misidentified until DNA testing in 2007 reclassified it as a fungus and it was then named Baudoinia after the French scientist who originally discovered the mould. Today, it is often referred to as whisky mould and is causing the global whisky industry considerable financial losses as new equipment has to be installed to lessen the effects of ethanol escaping into the atmosphere and feeding the naturally occurring fungal spores.

However, back in 1898 when Major James Grant, owner of the Glen Grant Distillery, returned from an African hunting trip accompanied with a new and undoubtably bewildered servant boy known as Byeway, no connection had been made between the black mould that grew in the locality and the whisky industry.

The African trip, which Grant had taken with his wife Fanny, had been a safari in search of wild game, and it was in Bulawayo, now Zimbabwe's second largest city, that he and his hunting party came across two skinny orphan boys who were accompanied by a cattle herder and his wife. The elderly couple explained to a multilingual member of Grant's party that they had found the two young boys wandering in the bush, and it was believed that they had somehow become separated from their own people. The two boys were believed to have been suffering from some sort of shock and neither could say where they had come from.

The story goes that the hunting party felt the boys' chance of survival were slim if they remained with the elderly, obviously destitute African couple, and so it was decided that one of the hunters would take the younger boy back home with him to act as a servant, while Grant took the other. Thus, without any legal niceties the adoption was arranged there and then. No doubt money or valuable goods were handed over to the elderly African couple, and thus the African boy named Biawa Byeway Makalanga found himself crossing a vast ocean and ending up in the small Scottish village of Rothes.

Little is known of Byeway's early life but his name Biawa seems to have been a corruption of the Shona word, *beewa*, which means stolen. This suggests that the young boy and his companion, likely his brother, may have been stolen from their original parents. This was a commonplace occurrence in Africa at the time, with tribes of nomadic people often warring with each other. Regardless of these early hardships, the young boy would go on to live a remarkable and fulfilling life. He would tell people that he was extremely grateful of the opportunities Major Grant gave him and he settled quite easily into the Scottish community where he now found himself. Initially he would serve as a pageboy and would attend the local school dressed in a Little Lord Fauntleroy-like outfit. It was not long before he was speaking English with a soft Rothes accent and became a popular member of the town.

When the First World War broke out Byeway found himself conscripted and he served with the Northamptonshire Regiment, eventually seeing action in the Middle East. He was awarded both the British War Medal and Victory Medal for bravery in battle, and when demobbed he returned to the Glen Grant Distillery and resumed his duties.

Byeway then joined the local amateur football team, where he gained a reputation as a solid goalkeeper and it was joked that he was the only African butler to ever play for a Scottish team. When

An elderly and distinguished-looking Byeway. *Author's collection*

his playing days were over, he remained loyal to his team and would always attend home matches and cheer on the boys. So popular was he that when he attended an away game, he would travel with the team on the team bus. He was treated like a VIP.

In 1931, Major Grant passed away and left provisions in his will that ensured Byeway would always have a home and a job at his distillery. At first Byeway was given rooms in the Glen Grant house and continued to work in the distillery and had his meals provided for him in the nearby Seafield Arms, but when the Second World War broke out the building was requisitioned by the 51st Highland Division and Byeway then took a job as a servant at a house in Lossiemouth. By all accounts Byeway didn't enjoy the experience and soon after the war he returned to take a room at the Glen Grant house, which by this time had been converted into flats for distillery workers. Byeway himself would pass away in 1972 and although his actual age was unknown, he was likely somewhere around 80.

Several years later a new still house was opened at the Glenrothes Distillery and from then on workers at the still would report sightings of the ghost of the African man standing by the new stills. These reports were taken seriously enough that in 1981 Cedric Wilson, a professor of pharmacology at Dublin University and an expert in paranormal activity, was called in to investigate. It was reported that when the professor first visited the distillery, he walked across the street to the graveyard and found himself standing next to the blackened stone that bore the inscription 'Byeway'. He was said to have spent some time conversing with the spirit at the graveside and then he announced that the erection of the new stills house had damaged lay lines that ran beneath the ground and this was making Byeway's spirit restless.

After using a dowsing technique to survey the ground, the professor oversaw two lengths of pig iron driven into the ground, each side of the new still house, and announced that these had repaired the damage to the lay lines that had been troubling the spirit of the African man. From that day to this the ghost of Byeway has not been seen, but the tradition of toasting the ghost continues at the distillery as each new batch of whisky is tested.

Scotland itself is a land of mystery, with ancient castles and misty valleys, and Glenrothes is far from the only distillery with paranormal legends connected. The Jura Distillery on Scotland's west coast is said to be haunted by the spirit of a vengeful old lady. The story is that in 1781 Laird Archibald Campbell, a supporter of the temperance movement, outlawed distilling on the island of Jura, and then some years later was horrified to be awoken in the night by the ghost of an old woman, who screamed at him for destroying the income of many of the inhabitants of the island. So terrified was the laird that he reversed the ban on distilling and later in 1810 he founded the Jura Distillery. A bottle of Jura, designed to appease the ghostly old woman, remains buried at the site of the original distillery. The ghostly story has inspired two official bottlings, with both Jura Prophecy and Jura Superstition being inspired by the legend.

The Highland Park Distillery in Kirkwall is often said to be visited by the ghost of its original founder, Magnus Eunson, and the Bowmore Distillery, on the Isle of Islay, has over the years been visited by the ghost of a headless horseman. However, by far one of the most terrifying of whisky ghost stories is the legend of the four-legged ghost said to haunt the Tomatin Distillery, close to Inverness. For centuries there have been reports in the area of the ghostly dog stalking people and then vanishing into a puff of black smoke. One of the most told stories of the spectral hound tells of a Tomatin worker who was walking home one night when he came upon the fierce hound. Despite his fear, he stood still and the hound approached him and sniffed at his feet, all the while a terrible growl coming from deep within its throat. The man was terrified but reached out and patted the dog, hoping to calm the beast. However, as soon as the man touched the beast it dissolved into a cloud of blue smoke and drifted off into the night. One can only wonder if the worker had been partaking in a little too much of the whisky at the distillery before making his journey home.

So well-known is the legend that the distillery has released a lightly peated smoky whisky called Cu Bocan, which in Gaelic means 'ghost dog', and Arthur Conan Doyle, perhaps Edinburgh's most famous son, was aware of the story and combined elements of it with the legend of a spectral hound that haunted Dartmoor in Devonshire when he wrote his Sherlock Holmes novel *The Hound of the Baskervilles*.

Feline Friends

Historically distilleries with their grain stores have had problems with rodents and it is no surprise that most distilleries kept a resident cat to help keep the pests under control, though with the near extinction of floor maltings and strict food hygiene regulations modern distilleries have seen their resident cats take on more of an ambassadorial role; there to greet visitors rather than exterminate mice.

Towser, the most famous distillery cat of them all. *Courtesy Glenturret Distillery*

One such cat who has taken on legendary status was Towser, who at the Glenturret Distillery, home of The Famous Grouse, is credited in the *Guinness Book of Records* with having despatched an incredible 28,288 rodents over a twenty-four-year period, which works out on average of three mice every day. These figures were calculated when a team from the *Guinness Book of Records* observed Towser's activities for several days and then used a statistical technique to come up with the mind-boggling figure of mice kills.

Towser, a long-haired tortoiseshell, was born on 21 April 1963, and up until her passing in 1987 she lived in the distillery, which is situated among the stunning countryside of Perthshire. Even today, decades after her passing, she is an important part of the distillery, with a bronze statue of the ferocious feline welcoming visitors. The cat has become a legend in the whisky industry, standing tall among many other famous distillery cats. Towser, by all accounts a slightly feral and bad-tempered cat, is still spoken about in awe by older distillery workers, who can recall the days when she patrolled the grain stores in search of yet another rodent to add to her impressive tally. The average lifespan of a domestic cat is between twelve and eighteen years, and yet Towser lived an incredible twenty-four years. Glenturret claims that her longevity is partly due to the fact that the stillman, whose job it was to care for the cat, gave her a daily nip of whisky with her morning milk. In 2017 Towser was further immortalised when Glenturret released a 13-year-old malt that went by the name of the Towser Cask.

Another famed feline went by the name of Passport and the story of her journey to the Glen Keith distillery is truly incredible. It all stems from the United States 1938 ruling that strictly states bourbon must be aged in new American oak barrels, which means that once the barrels have been used they are surplus to requirements. Now Scotch whisky has no such ruling and even today the majority is aged in previously used barrels, which can often have a favourable impact on the taste of the finished spirit. A Scotch aged in former sherry casks, for instance, will have a distinctively different profile to one of that aged in former bourbon casks. This is why Scotch producers often make a big deal of the previous use of their barrels in their marketing.

It is also cost-effective for Scotch producers to import previously used barrels and although Scotch is often aged in former sherry, port or even wine barrels, the great majority of those used are those that once held bourbon. Whisky, like all spirits, is clear when it comes from the still and it gets much of its colour and a great deal of its aroma and flavour from the barrel used in the ageing process. Many Scotch enthusiasts simply prefer their drink to have been aged in ex-bourbon barrels. It is a truly symbiotic relationship between bourbon and Scotch that has existed for a great many years.

In 1993, at the Glen Keith warehouse in Speyside, a containment of sealed ex-bourbon barrels had just arrived from Kentucky when one of the workers detected a faint scratching sound coming from within one of them. Alan Greig, director of heritage at Chivas Brothers, the parent company of Glen Keith, soon received a phone call from a worker at the distillery. He contacted the Scottish animal welfare charity the SSPCA and was alarmingly advised to drill a hole in the barrel and gas whatever critter had somehow got itself trapped inside. No one had any idea of what sort of critter it was, and although the most likely scenario was that it was some kind of rodent, a rat maybe, the warehouse workers, unsure of what species were prevalent in Kentucky, all made suggestions.

'A racoon maybe,' said one.

'What if it's a skunk? asked another.

It was decided to open the barrel and try and tempt whatever was out with a tray of food. Everyone was shocked when a small black and white kitten crawled out and stumbled around the warehouse floor. The kitten was immediately named Dizzy, given that she was so unsteady on her feet, which was not surprising given that she had somehow survived four weeks within a whisky barrel with only condensation collected on the wood to provide sustenance.

The cat was a massive hit with distillery workers and she soon became chief cat at Glen Keith distillery, where she was renamed

Glen Keith, the home of Passport the globetrotting feline. *Author's collection*

Passport. It was a fitting name given her incredible world travels but also because she had found her new home at a distillery famed for its blended whisky called Passport. She remained at the distillery until 1999 when the distillery closed down and after that she moved to the Strathisla Distillery. She was soon taken home by a distillery worker and then lived out a long and happy life. It was never discovered which Kentucky distillery she had come from, and all attempts to find out failed, but wherever it was her story is truly wonderous.

These days it is something of a tradition for distilleries to have a resident cat or two, but most of these live pampered lives, spoilt rotten by both those who work at the distilleries and visitors who make the pilgrimage to learn more about their favourite tipples. Mice and rats are no longer the problem they once were, but in years gone by distilleries would use a process called floor malting where the unmalted grain was spread on the distillery floor and then turned by hand until it was ready for the kiln. This obviously meant a lot of waste, which was a magnet to rodents, but now that most modern distilleries outsource their grains from malting companies, distillery cats tend to serve a strictly ambassadorial role, simply being there to charm the visitors and pose for photographs. These snapshots are more often than not shared on social media, providing free advertising for the distilleries in question.

Scotland's Cannibal Clan

When a story is colourful enough that it becomes legend, then it doesn't really matter that it was likely born of half-truths, exaggerations and barefaced lies, such is the case with the story of Scotland's Sawney Bean, illicit whisky maker, smuggler and cannibal.

In the nineteenth century, readers thrilled at the lurid stories published in the *Newgate Calendar*. The bulletin, first published in 1824 and continuing until 1826, was a mix of fact and sensational fiction. It was hugely popular with readers and although treated with disdain by many it was hugely influential and is mentioned several times in Dickens' *Oliver Twist*. The title was originally published as a straightforward list of executions produced by the keepers of Newgate Prison, but the title was soon used by other publishers, who put out a series of cheap books about notorious criminals such as Dick Turpin, John Wilkes and, of course, Sawney Bean.

> *The following narrative presents a picture of human barbarity, that were it not attested by the most unquestionable historical evidence, would be rejected as altogether fabulous and incredible.*
>
> The Newgate Calendar

It is from these publications that much of the legend of Sawney Bean derives and although he had previously been mentioned in several notable publications, actual documented evidence is non-existent. There is one school of thought that claims the original story was written by Daniel Defoe, author of *Robinson Crusoe*, in order to denigrate the Scots after the Jacobite uprising. Defoe had been in Scotland as a spy and propagandist during the 1707 parliamentary union and many believe he held the pen behind the tales that originally inspired the *Newgate Calendar*. He is known to have used almost 200 pen names, which was common practice during the period but there is no way to confirm his authorship of the Sawney Bean story.

The accepted story is that Alexander Bean was born in East Lothian sometime during the sixteenth century, and he later became known as Sawney – the name Sawney was an English derogatory nickname for a Scotsman, much in the way we use the name Jock today. This fact alone seems to add credence to the belief that the story of Bean was actually an invention designed to demonise the Scottish race, but it matters not because today much is made of the tourism possibilities of the tale. Many tourist maps, including the ordnance survey one, show the location of Sawney Bean's cave, so if it was all a fiction then it seems to have been given the trappings of historical fact; as Carleton Young says in the John Ford movie *The Man Who Shot Liberty Valance*, 'When legend becomes fact, print the legend.'

Bean is said to have left home when a young man and soon met up and married a woman called Agnes Douglas, but known as Black Agnes. The two of them set up home in a cave system at Bennane near Ballantrae in Ayrshire, Scotland. The reason the newlyweds decided to populate the caves was totally logical, given that neither of them wanted an honest trade and decided to make a living from robbery, which soon turned to murder and then cannibalism. They would ambush travellers on the narrow roads above the caves, and in order that they would never be identified it was decided to kill their victims – dead men after all tell no tales. From there it was just a short stretch to actually eating their victims, preserving the flesh in barrels of the moonshine whisky that Sawney produced. The high-protein diet seemed to do the trick and soon Black Agnes gave birth; it is said she produced fourteen children in all and as these children matured they incestuously produced babies themselves until the Bean clan numbered more than forty in total. The entire family did well on a diet of rich meat and illicit whisky as their tally of victims increased.

Reports were often made over the years that many grisly but well-preserved body parts were washed up on the surrounding beaches, and although mass searches were carried out by the authorities, nobody ever thought to explore the remote cave system, the entrance to which was flooded twice daily during high tide. An idea that nothing human could survive in the caves seems to have been the thinking of the authorities. If one visits the caves today then such an attitude is clearly understandable.

The cave system is situated on the Ayrshire coast and the entrance near the waterline can only be reached on foot and with great care. After descending 150ft from the car park on the A77 road at Bennane Head, which is 8 miles north of Ballantrae, visitors have to scramble over a large, nay enormous, boulder to reach the cave entrance, but nevertheless many thousands of people make the pilgrimage each year. The Sawney Bean story, whether truth, legend or brazen invention, is certainly a money-spinner for this treacherous though stunning beauty spot and its surrounding areas.

> *Sawney and his wife took shelter in this cave, and commenced their depredations. To prevent the possibility of detection they murdered every person that they robbed. Destitute also of obtaining any other food, they resolved to live upon human flesh. Accordingly, when they had murdered any man, woman or child they carried them*

> *to their den, quartered them, salted and pickled the members and dried them for food.*
>
> The Newgate Calendar

One evening in 1599, the legend tells us that a man and a woman were riding home from a fair, sharing the same horse, when they were confronted and attacked by the Bean clan. The woman was promptly pulled from the horse and had her stomach torn open and her entrails pulled out while she screamed her last. Her horrified husband, having witnessed the horror, rode his horse at the Bean clan, trampling several of them underfoot.

'Fiends,' the man screamed from horseback. 'Inhuman fiends. I'll kill every one of you.'

The man happened to be armed and he fired at the ravenous Beans and the commotion soon attracted more than thirty people who were travelling home from the same fair that the man and his unfortunate wife had been returning from. The Beans, for the first time finding themselves at a numerical disadvantage, fled into the woodland and made their way back to their cave, where they assumed they would be safe.

The grieving husband was taken to Glasgow, where he told his incredible story to the chief magistrate. A connection was soon made between this horror and the ever-growing list of unexplained disappearances in the area that had been a continuing mystery for almost a quarter of a century, and it was decided to take the matter straight to King James VI.

The King, along with 400 men and a pack of bloodhounds, soon arrived in the area and undertook an extensive search of the surroundings. This time the cave at Bennane was not overlooked thanks to the bloodhounds, who went crazy when taken near to the partly submerged cave entrance. When the tide had retreated, the dogs along with several men entered the caves and found a scene that was reported to have been nightmarish. The Bean clan, some of them still sporting injuries from their recent battle with the man on horseback, were discovered along with numerous human remains: there were body parts hanging from the walls, other parts were preserved in barrels of moonshine whisky and there was a mound of clothing, jewellery and other items that had been stolen from the unfortunate victims.

The legend goes that Sawney Bean himself, along with all the male members of his family, were publicly executed by having their limbs cut off and being allowed to bleed to death. The woman members of the Bean clan, after watching their menfolk die, were then burned upon a series of large fires.

True or not, the legend of the whisky-making cannibal is certainly a colourful story and today visitors to the Edinburgh dungeon can be scared witless by the show put on in the exhibit that details the Sawney Bean legend.

Let's raise a glass to the cannibal.

No Place for the Devil to Hide

Bowmore, the devil's cask. *Author's collection*

The Kilarrow Parish Church, better known as the Bowmore Round Church or Round Kirk, at Bowmore on the Isle of Islay, is a remarkable building. It was designed in 1767 by Daniel Campbell and inspired by the Italian churches he had seen on his many travels across Europe. The church, first opened in 1769, is still being used for services to this day.

Legends surround the church, the most famous being that the devil himself visited one winter's evening in 1837, but because the church was perfectly round there were no corners in which the devil could conceal himself. It is said that the congregation of the church chased the devil from the church and that the demonic being fled to the nearby Bowmore distillery never to be seen again. That same night a small paddle steamer, *The Maid of Islay*, set sail to transport barrels of Bowmore whisky to the mainland. It is said that the devil had concealed himself in one of the casks of whisky only to return to his demonic realm when able. So well known is the legend that in 2021 Bowmore released a special 23-year-old whisky called No Corners to Hide. Only 666 bottles were produced – those marketing men again – with label artwork by Frank Quitely, who had honed his art of creating demonic imagery in his work for both Marvel and DC comics.

The Great Bourbon Rip-Off

When demand outweighs supply then things can get silly in the world of whisky, with bottles of rare expressions going for vastly inflated, even eye-watering, sums in a thriving secondary market. No wonder then, that whisky often catches the eye of the unscrupulous and downright crooked.

In 2013 a story made international news when the Buffalo Trace Distillery in Kentucky announced that in excess of 200 bottles of its ultra rare Pappy Van Winkle had gone missing, vanished, spirited away. There are only around 8,000 cases produced a year, less than a thousandth of the output of say Jim Beam and a mere fraction of that churned out by Jack Daniel's, and the expression is one that regularly goes for silly money on the secondary market, with insane prices of several thousands of dollars a bottle not being uncommon. The whiskey is known more for its high pricing rather than its quality, because so few people are ever in a position to taste it. Those bottles that are released sell out instantly, even when priced at between $100 and $300 a bottle. It is then that some of these bottles reach the secondary market and, as noted earlier, the prices become quite absurd. When millionaires who can buy anything suddenly find there is something they can't buy they get overexcited.

Pappy Van Winkle is a long-aged expression from the Buffalo Trace Distillery that often spends more than two decades in the barrel, so over the years the barrel loses a sizeable share of its contents to evaporation, the so-called angel's share, and after two or more decades a 55-gallon barrel, given the dry Kentucky summers, can lose around 50 gallons, which leaves only 5 gallons for bottling. This is why the 23-year-old Pappy Van Winkle, which often leaves tasting experts wildly spouting adjectives to describe its sublime taste, is so highly sought after.

> *We make fine bourbon, at a profit if we can,*
> *At a loss if we must, but always fine bourbon.*
> The Pappy Van Winkle motto

Those that can afford the whiskey rarely drink it and instead keep it as an investment, which is basically an absurdity in itself. Over the years, I've met many whisky collectors and seen some impressive collections but I don't look on them in awe. In truth, it leaves me cold for I can't fathom the attitude and I fail to grasp the point. Surely the sole purpose of whisky is for it to be drunk and not displayed in some fancy cabinet! Oh, well each to their own. Horses for courses, and all that. Still, a bottle of whisky or whiskey, valued at thousands, sitting on a shelf seems so pointless, so sad. It's all beyond me, and even empty bottles of the 23-year-old Pappy can fetch more than $100 on eBay. The lunatics really have taken over the asylum.

Gilbert Curtsinger. *Franklin County Police Department*

With that kind of money in play, then it would seem inevitable that there would be thefts and in 2015 it all came to light when Gilbert Toby Curtsinger was arrested and charged with bourbon theft on a grand scale. The 2013 theft of Pappy Van Winkle was also laid at his door, but he denied any involvement in that particular crime and no clear evidence was ever discovered to tie him to that theft.

> *I was the kind of guy that you knew that could get you stuff,*
> Curtsinger to a 2022 Netflix documentary that looked at his bourbon heist, which had become to be known as Pappygate

An anonymous tip-off had led the police to Curtsinger, but when they arrived at his property all they were able to find were two barrels of whiskey in his shed, and later analysis revealed that the whiskey was not Pappy Van Winkle, nor any of the other whiskeys made at the Buffalo Trace Distillery, but rather Wild Turkey, a distillery that had not reported any thefts whatsoever. However, Sherrif Pat Melton of Franklin County felt he had enough evidence to charge Curtsinger, and a later search would uncover three further whiskey barrels containing Wild Turkey, as well as twenty-five bottles of Pappy Van Winkle. The police now had their man and eventually even more barrels were discovered, and court documents reveal that in total eighteen barrels, containing both Wild Turkey and Buffalo Trace, had been retrieved.

Curtsinger had been an employee at the Buffalo Trace Distillery in Frankfort, Kentucky, for twenty-six years by the time of his arrest. He told police that staff would often take a bottle or two home, that it still went on and was regarded a perk of the job rather than theft, but nevertheless he admitted to the theft but denied that he was part of a larger operation. In all ten people were arrested, including Curtsinger's wife, Julie. Despite no real evidence Curtsinger was painted as the ringleader in a syndicate that included several other Buffalo Trace employees, a recreational softball team and Curtsinger's own father.

It is likely that this image is the closest that most of us will get to a bottle of Pappy. *Author's collection*

The total thefts were estimated at valuing around $100,000 between 2008 and 2015.

Out of the ten arrested, Curtsinger was the only one to face jail time, having been sentenced to fifteen years in 2018. But given that the evidence against him was so slight he ended up serving only thirty days and was released on a system called shock probation, in which the convicted felon serves out the rest of their sentence under licence. This would seem a fair ruling given that, although Curtsinger had been sentenced as the ring leader of a syndicate, there was little evidence that the thefts had been organised. In actual fact they seemed anything but, and a great deal of the whiskey recovered had not even been reported as stolen. While it was true that Curtsinger had stolen whiskey over a long period of time, he certainly wasn't the only one doing so and as the authorities were unable to bring charges against anyone else, then the case against him was decidedly weak. The fifteen years' prison servitude was looking unduly severe given that Curtsinger was of previous good character.

The truth behind the so called Pappygate saga is that this was not one theft, but a series of robberies that had been going on for years and Curtsinger, who will always be known as the Bourbon King thanks to the Netflix documentary, was no ringleader. He was just a part of a much larger string of thefts, mostly driven by opportunity, that went back for many years and his so-called syndicate was merely a group of friends who would deliver the whiskey at a knock-down price to grateful people in the community. The Kentucky distilleries involved in the thefts kept their distance from the documentary makers, likely embarrassed by the theft culture that seemed endemic in the industry. At the time of writing the 2013 theft of all that Pappy Van Winkle has never been truly solved and it is unlikely that it ever will be.

'Psst, anyone after a cheap bottle of Pappy?'

The result of all this was that Buffalo Trace and other distilleries made significant moves to tighten up their security, Curtsinger and others obviously lost their jobs, and a fair few unknown people received bottles of Pappy Van Winkle and other expensive bourbons at a knock-down price.

Chapter 5

The American Experience

The origins of whiskey in America can be traced back to the initial European settlement of the continent itself. When the English settlers arrived in Jamestown in 1607, they had brought with them materials to build shelter for themselves and their livestock, as well as food stuff, seeds for planting and a vast supply of ale.

A significant portion of the holds upon the ships that had transported the settlers had been allocated to the storage of ale, which was important because drinking water was often scarce during the Atlantic crossing and ale would provide much-needed calories for those who made the journey. The ships that reached Jamestown had gone by an indirect route that had taken them to Africa and then down through the Caribbean before reaching what is now North America. This rather wayward route was taken because it provided stops to restock supplies, but even then stocks would get low and the ale was seen as a necessity rather than a luxury. The ships had left London in December 1606 and finally after months at sea reached Jamestown in May 1607.

Of course, it would be several more decades before whiskey was first produced in America, but the early settlers started brewing ale almost immediately, and in 1609 it was recorded that a field had been set aside for the cultivation of barley to be used in the production of ale and a request was sent to England for several brewers to be sent over during the next supply run to the settlement. Many men took up this offer and Jamestown became the site of America's first brewery, the remains of which are now preserved at the Jamestown Historical

Site. It must be stated, however, that this was not America's first commercial brewery; that distinction goes to the New Amsterdam, now New York, Block and Christiansen Brewery, which was set up in the Dutch settlement in order to provide ale for the seamen, merchants and traders who were passing through with an almighty thirst.

The brewery in Jamestown then was not a commercial operation but was set up to supply its own settlers with ale, but the brewing of ales largely became the duty of the much put upon housewives of the era. *The English Huswife*, a book published in 1615, became a bestseller and many copies were sent across to Jamestown, its pages filled with recipes for Banbury cake, gingerbread and other treats as well as a full chapter dedicated to the art of brewing ales at home.

As populations grew and corn fields became more abundant, farmers would begin trading grain with their townsmen but eventually there was more grain than was needed and farmers had the problems of storing all that they had produced. Making ale was still highly popular, but this took a lot of time as well as storage space and there were some settlers, particularly those from Scotland and Ireland, who had brought small stills with them when they made their journey to their new lives. It was only logical to use the surplus grain to make corn-based liquors that could be stored for many years without the risk of spoiling, and this was America's first form of the drink which would eventually evolve to become the whiskey that is today enjoyed on a global scale.

Jamestown, though, was only the start of it and soon the expansion would begin. In 1767, Daniel Boone was exploring the western wilderness, and telling stories of fertile soil, vast wide-open ranges and hunting grounds that were second to none. Many were attracted to the promise of such fruitful lands in which to forge their destinies, and although Native American populations stood in the way of this expansion, this was not seen as a hurdle, and the tribes that would not sell their land to the new settlers were quickly driven away by American militia. The European settlers believed that they had a divine right to settle in North America, taking the land from the indigenous people by force if necessary and this would lead many years later to the policy of Manifest Destiny being adopted by the fledgling American government. The phrase Manifest Destiny was first coined in 1845 and it meant that the people of North America were destined by God to expand their dominion and spread democracy and capitalism across the entire continent. The Americans would use this philosophy to justify the forced removal of Native American populations.

New farmlands opened up, and settlers began to arrive from, among other places, Scotland and Ireland. Many of these people were fleeing problems such as drought, food shortages and religious persecution in their respective homelands and America seemed to promise them new opportunities and ever-expanding lands. More than a quarter of a million people immigrated to America during the 1700s, and initially many of these established themselves in Pennsylvania, but from there a great deal pushed west in search of cheap land upon which to stake a claim. This started the great westward expansion that would continue for many years, give the concept of Manifest Destiny to the American people and reshape the entire continent.

It was in 1776, with the American War of Independence in full swing, when Kentucky County was officially established. Initially it was a part of Western Virginia, and the governors of that state wanted to encourage the drive into the fertile lands that had yet to be claimed. A new law, the Virginia Land Act, was enacted that enabled settlers to claim 400 acres of land if they built a home and planted at least one corn patch before 1778. Again, the pre-existing ownership by the Native American tribes was ignored out of hand.

Immigrants quickly took advantage of the act and claimed land and, of course, many of these newcomers were skilled workers who knew how to farm, produce good-quality crops and were also knowledgeable about the art of distillation. Soon distillation was being practised all across this new frontier, and as westward expansion continued crudely produced spirits went with it.

Whiskey had gained a foothold in America.

One of America's finest. *Author's collection*

The Scottish and Irish immigrants had primarily been making rye whiskey because the European rye crops they had brought with them were more durable to the new climate than the barley they were accustomed to using back home. Barley was a far more temperamental crop that needed a much more careful cultivation than the rye, which seemed to grow in the Kentucky soil with the temerity of a weed, and soon rye whiskey became synonymous with the newly formed Kentucky County. To this day Kentucky remains the state that produces the most whiskey in the United States. To jump forward a few hundred years for a moment, it is worth pointing out that although Kentucky remains synonymous with whiskey and both Buffalo Trace and Wild Turkey, which rank among my favourite bourbons, hail from the state, production is practised across America. Tennessee comes a close second to Kentucky, Jack Daniel's likely being its most famed brand, but there are good-quality spirits coming from Wyoming, Idaho, Montana and even Seattle, Washington.

Today American whiskeys are far more than rye based. In fact, predominately rye-based blends are in the minority and, as well as bourbon, we have American single malts, blends and everything in between. However, deep within the genome of all of these whiskeys is the nucleus that started with the immigrants who brought their craft from Scotland, Ireland and far beyond.

Chapter 6

Old Glory, Independence and Whiskey

In 1776, at the start of its war for independence, the American colonists found themselves in crisis because, while Great Britain was hamstrung by the time and effort it took to send troops and supplies to the far-flung battlefields, the Americans were limited in their ability to effectively wage war against their overlords due to having a smaller population, a shortage of weapons and a chronic lack of supplies. Food shortages saw riots break out between 1776 and 1779, and it became evident that a shortage of food could do more to destroy the morale of regiments than any victories over the British could ever do to rebuild it.

The continental armies had to face the fact that getting food supplies to their men was not at all easy. The planting and growing of grain took time and that was before the harvesting, milling and transportation was taken into account. Flour and bread would spoil, often being infested with worms and weevils during the journey to the troops. The provision of good-quality meat was also a problem since wild animals were obviously scarce when there were scores of marching soldiers in the vicinity and the time it took to raise suitable animals on farms presented its own problems. Much of the meat provided to the armies was dried and preserved in salt but even then some of it was spoiled in transit, which meant that the rations supplied to soldiers was severely limited. It was clear that the continental army could not function if it could not keep its troops fed.

There was also a shortage of antiseptic to treat wounded soldiers and soon the American military leader George Washington realised that liquor might provide the answer to some of the problems. Rum could be used to provide much-needed calories for the soldiers, as well as help combat the terror of battle. It could also be used as a field antiseptic and soon it became to be regarded as essential if there was any hope of beating the British armies who were stepping up in their attempt to put the Americans back in their place. The liquor travelled well but when the British started cutting off naval trade routes it became a problem to supply enough to the troops.

> *The benefits from the usage of liquor have been experienced by all armies.*
>
> George Washington to John Hancock,
> the president of Congress, in 1775

The uncomfortable fact was that rum was becoming increasingly difficult to come by. The British blockade was stopping the importation of the raw materials needed to make the drink, but the essential crops to make whiskey were plentiful and like rum it could be used as an antiseptic, as well as keeping cookware sanitary when used as a wash. More importantly it could be given to the troops as a source of sustenance as well as in helping to stave off the numbing cold of winter. Washington ordered that each soldier be provided with one gill, 4oz, of whiskey a day as part of their rations, and also directed commanders to issue extra rations of the drink as a reward for valour in battle. Washington also requested that Congress fund the erection of new distilleries and although this was rejected on monetary grounds it was not long before enterprising businesses started to pop up in order to supply the demands of the continental army. These distilleries were situated far away from the main theatres of battle and escaped interruption from the British forces.

The whiskey industry in Kentucky, Pennsylvania, Virginia and Maryland benefited hugely from Washington's policy – a gill a day equated to almost a bottle a week for each and every soldier, tens of thousands of bottles a year, which was a massive market for the then nascent whiskey industry. That year alone the army purchased more than 70,000 gallons of whiskey for the use of its troops.

Whiskey was seen as being of great benefit to the troops. After all, way back in the sixteenth century, Irish alchemist Richard Stanyhurst had extolled the curative properties of the drink:

> *It beying moderatellie taken does sloweth age, strengthen youth, relishes the heart, lightens the mind and quickens the spirits.*

He also recommended the drink for curing dropsy, resolving kidney stones and intestinal gases, and declared it as being good for the circulatory system. Whatever the truth of such claims, it was a simple fact that drinking whiskey was safer than consuming the filthy water often provided to the troops.

Whiskey soon became an essential part of army life and sutlers – retailers licensed by the military – set up camps around military installations and would do a roaring trade selling tobacco, candy and whiskey. These traders were allowed to supply soldiers on credit with costs being recouped from the man's wages when the bill was presented to the troop commanders. Each soldier may have been entitled to one gill a day, but the greater majority were drinking far more thanks to the sutler camps that followed them around battle to battle.

When the war ended with a hard-fought victory over the British in 1783, soldiers slowly started to return home to their families and found themselves continuing to enjoy the whiskey they had become accustomed to from the long years on the battlefields. The distilleries, many of them started up to supply the wartime demand, now found themselves answering a consumer need for their spirits that was as voracious as ever.

The whiskey market enjoyed a boom but trouble was not far away and in 1789, when George Washington became the first president of the United States, the newly formed government had to find a way to pay off the significant debts it had incurred during the war. They would eventually turn to the distilleries, those same distilleries that had sprung up to help them with rations during the war, and passed a whiskey tax. This was the first tax to be placed on domestic goods since the founding of the United States and would provoke a rebellion.

Also called the whiskey insurrection, the rebellion was an often-violent uprising that occurred in the new country between 1791 and 1794. The hard-fought revolution against the British had left America with crippling money problems; the national debt was estimated at $54 million; a vast sum even today but astronomical during that period. To further alarm the newly formed American government, there were also the individual state debts, which collectively added an additional $25 million. The Secretary of the Treasury Alexander Hamilton urged Congress to consolidate the debts into one figure that would be managed by the federal government, which would allow for taxation to not only reduce the deficit but also provide a financial system that would, at least in theory, promote national unity and American prosperity. Congress, with some reluctance, approved these measures in July 1790, and Hamilton set about figuring out a workable form of taxation that could ease the burden on the finances of the developing nation.

America's first excise tax was placed on distilled spirits and although many in Congress were opposed to the new tax being placed on a domestic product, calling it the leading edge of tyranny, most took the view that because liquor was a luxury good, rather than a basic necessity, then paying the levy was a choice that citizens could make. Washington was weary of the idea himself; he was deeply aware of the growing divide between the eastern majority and the tiny western minority and didn't want to introduce a tax that could be seen as targeting western farmers while eastern businesses remained largely unscathed. However, after taking a trip to Virginia and Pennsylvania and speaking to officials he found that the tax proposals had some support, although he would soon realise that the views of a limited

Alexander Hamilton in a portrait by John Trumbull. *Getty images*

number of officials were not necessarily reflective of those held by the greater populace. Nevertheless, Washington felt empowered by the views of the officials in the more rural areas and he promptly returned to the capital and signed off on Hamilton's tax proposals.

The tax was actually imposed on all distilled spirits, but because the consumption of whiskey was so widespread it soon became known as the whiskey tax. The Excise Whiskey Tax Act of March 1791 meant that spirits were taxed by both volume and strength. This extra cost would be passed on to the consumer, with it factored into the price of each individual bottle or drink sold. There was also an additional annual tax on stills based on the amount of alcohol they could produce. The revenue the government could expect to receive from such taxation would be sizeable and help achieve the aims of the founding fathers, which was for the entire nation to become self-sufficient.

This was, Hamilton argued, the least objectionable form of taxation the government could impose on its citizens, and he gained much support from social reformers, who felt this was a sin tax that would help people realise the harmful effects of alcohol. However, things would not play out as expected and right from the off the tax proved controversial. For a start, the way the system worked was regressive – basically because of the way the legislation was set up it meant that the more whiskey a distillery produced the less tax it would end up paying. This meant that the large whiskey producers were able to cope but the smaller operations, largely carried out as a sideline by farmers, were faced with enormous bills.

This was especially galling given the limited product these small distillers actually produced. Another bone of contention was that it was a legal requirement to register every still, with enormous fines given to those who failed to do so. Many people in the rural river valleys owned and operated their own stills, which they would use to produce whiskey for their own consumption, but they would also use it to barter for other goods without any actual cash exchanges taking place. All of these people were required to register and ultimately pay tax on their stills.

Society, at this point in history, had not entirely or even marginally switched to a cash-based trade system and bartering remained a substantial cultural norm, especially for those in rural areas. To these people the bartering system was essential to their very survival and now under the new system they were going to be taxed on a product that brought them very little, if any, financial gain in the first place. Many people felt that the fledgling government had become no better than those they had fought against in the war for independence. It must have seemed to many that all the long struggle for independence had actually achieved was in replacing King George III with an American version. It was an incendiary situation and further fuel was poured on smouldering fires when it was announced that tax evaders would be tried in state rather than local courts. The journey to state courts was beyond the means of many of those affected by the tax, and although the rules would eventually change in favour of local courts it was too little, too late. The damage had already been done, and many who had fought against the British for the promise of freedom from political tyranny were feeling aggrieved with their new government.

Protests against the tax were immediate, with smaller producers furious at the unfair situation that allowed large operators to pay an annual tax of 6 cents per gallon with large tax breaks for the more whiskey that rolled off the stills, while the small operations were left paying 9 cents per gallon and the fact that tax payments had to be made in cash ignored the barter system entirely. Refusal to pay was common, as was the intimidation of officials hired to collect the taxes. The fact that the government reduced the excise levy the following year did little to defuse the anger of rebels.

Violence was inevitable and in September 1791 excise officer Robert Johnson was travelling through western Pennsylvania when he was set upon by a group of men dressed as women. The men dragged the terrified tax inspector from his horse, stripped him naked and then tarred and feathered him before leaving him afoot in the woods. Although not fatal, this was an extremely painful punishment and the hot pine tar poured over the body would blister the skin. Pine tar is produced by heating pine wood in a low-oxygen environment. This results in a glue-like, incredibly sticky substance that when removed would more often than not take a layer of skin with it.

The anger increased, particularly among those based on the western frontier, whose entire trade system was based outside of

A humiliating and incredibly painful experience. *Public Domain Image*

the American economic machine. These small farm-based distillers felt betrayed by a government that had encouraged them to produce whiskey for use by the military during the war. This was a major headache for the government and when the Indian Wars escalated towards the end of 1791 the government found themselves unable to protect settlers let alone collect on unpaid taxes.

It was during this period that the American moonshiner came into being. These men, like those before in Scotland and Ireland, started moving their whiskey under the cover of moonlight and struck up a vibrant trade away from the eyes of the authorities. More and more people started producing moonshine whiskey as a way of maintaining a way of life without government interference. And now tax collectors were refusing to do their jobs for fear of retribution from angry mobs. This was understandable as when a cattle drover named John Conner was sent out to arrest two of the men it was believed had been among those that had tarred and feathered Robert Johnson he himself suffered the same fate. He had been discovered several hours after the event bound to a tree. Following this event, it soon became public knowledge that Johnson, fearing further violence, had resigned from his position as a tax collector, which empowered the rebels and they vowed to continue their fight against the system.

Alexander Hamilton, though, was having none of it and he sent out a court official, this time with a small group of armed guards, to deliver the warrants. However, the angry mob remained defiant and quickly overpowered the men without a shot being fired. By the end of the day the court official and guards had been whipped, tarred and then feathered. This left Hamilton furious and George Washington at a loss as to how to proceed against the seeds of rebellion that were starting to threaten the unity of his fledgling country.

Earlier in the year the rebels, led by the moderately minded Hugh Brackenridge, had held a meeting at Redstone Old Fort in Pittsburgh and the take away had been that violence should be avoided at all costs, and that a team of delegates should be put together to take the grievances directly to the government. However, by the following year, with violence against tax inspectors now commonplace, a further meeting was held and this time things took a far more radical turn. The rebels set up their own extra-legal courts, which were independent of the state and federal legal system. They also strengthened their refusal to pay taxes and set about setting up a people's militia in order to protect the farmers and landowners from the federalists. The country was, in effect, at war with itself.

The opposition to the tax was now widespread, and as well as Pennsylvania, there were skirmishes between rebels and tax collectors in Maryland, Virginia, North and South Carolina and Georgia. It became clear that the government was losing its authority, that the rebellion was testing the constitution itself and both Washington and Hamilton feared that a military solution to the problem would have to be enacted. This was not something anyone in the government wanted to consider at the present time, with Attorney General Edmund Randolph arguing that there was not enough evidence to legally sanction such a response.

The rebels, though, were now well organised and newspaper articles written under the name Tom the Tinker began to appear in the *Pittsburgh Gazette*. These threatened those who compiled with the whiskey tax, and made the case for the rebels. The columns were sent directly to the editor, John Scull, together with instructions to print them in the next issue of the newspaper or else expect a visit from Tinker and his men, and such was the terror of the times that the requests were always promptly obeyed. Tinker is believed to have been a pseudonym of John Holcroft, an educated farmer and distiller, who as well as holding several public offices was one of the leading lights in the rebellion. Holcroft always denied that he was Tom the Tinker, even after Washington granted him, along with other rebellion leaders, full amnesty for their actions during the insurrection.

During the rebellion, the Tinker became a symbol of defiance and freedom and the rebels took to calling themselves Tinker's men and would yell, 'Hurrah for Tom' before confronting tax collectors. And confront them they did, with not only physical attacks on the tax collectors themselves, but also extortion of local officials while moonshine distilling continuing apace. Distillers who had complied with the taxes were also targeted and there were several instances of stills being destroyed and barns being burned down, which would further reduce government revenue.

It would all come to a head in 1794 when for the first time Washington led a troop of militia, some 13,000 strong, against his own citizens. This was made possible under the Militia Act of 1792, which stated that where the normal channels of law and order had broken down then a militia could be formed in order to restore a lawful situation.

The events that led to this started when the federal courts issued more than sixty subpoenas against distillers who had refused to pay taxes, and officials were sent out to deliver the writs. However, on 15 July things took a deadly turn. Federal Marshal David Lenox and General John Neville, recently having taken up the position of regional tax inspector, turned up at the farm of Oliver Miller in Allegheny County, Pennsylvania, in order to serve a writ but Miller refused to accept the papers and strongly suggested the two men leave his property. Lenox fired warning shots but, fearing that a mob would soon arrive to aid the farmer, the two men departed quickly without serving the subpoena.

The following day Miller and thirty other men surrounded Neville's home on Bower Hill, demanding the general surrender himself and face trial, but Neville fired at the crowd, fatally wounding Miller. A battle broke out between the mob and the defenders of Neville's home but the house was well fortified and eventually the crowd retreated. They returned the following day, this time 500-strong and led by Revolutionary War veteran James McFarlane. However, Neville also had reinforcements within his property, with a dozen soldiers from the nearby Fort Pitt having been sent to offer some protection to the tax inspector.

McFarlane and Neville exchanged dialogue but when it appeared inevitable that the situation was going to escalate, the women and children were allowed to leave the property. Almost immediately shots sounded out from both sides. McFarlane called a cease fire when he spotted a white flag in one of the windows of the house and, believing that Neville was going to surrender himself, he moved towards the house. However, a single shot from one of the windows then killed him instantly.

The rebels immediately responded with a ferocious wave of gunfire, peppering the house with lead shot as a group of men rushed forward and set fire to every structure. As the fire took hold in the house, Major Abraham Kirkpatrick and John Neville emerged and were taken prisoner. They would later escape from insurrectionist custody and would go into hiding until the rebellion was finally quashed by a huge show of force as Washington led his militia against the rebels.

In September, Washington marched into Pennsylvania, the first and only time that a sitting president has led troops into battle, but faced with overwhelming odds the rebels quickly dispersed. The president

Commemorating the spot where Neville's mansion stood. *Author's collection*

must have been relieved that no shots were fired, as being seen to lead troops against his own citizens could have stoked up even more anti-government feelings. This was to be avoided if possible because the government was aware of the consideration by some of the more moderate and well-placed rebels of seceding from the United States.

The whiskey rebellion was no more and although many arrests were made, only two men, John Mitchell and Philip Weigel, were found guilty of treason and sentenced to be hanged. However, they were later pardoned by the president when it became clear that carrying out their sentences would be counterproductive and martyr the men. While the violence may have stopped, the resistance to the whiskey tax was as strong as ever and illegal distillers continued to produce their product under the radar of the federal authorities. Many of them fled to Kentucky, where between 1794 and 1800 official records show that 177 illicit distillers were fined. However, this was just a small percentage of the illicit stills that operated in an area that was fast becoming the unofficial whiskey capital of the United States. Those that were fined simply paid up, shrugged their shoulders and moved their stills elsewhere.

The whiskey tax would eventually go, but it would take a new administration to repeal the law. Indeed, Thomas Jefferson successfully ran his political campaign on the promise of ending the levy. When he took up office in 1801 as the third president of the United States he quickly led Congress in repealing the tax and in its place implemented federal tariffs that were widely successful as US foreign trade began to grow.

Chapter 7

The Birth of Bourbon

Like the history of distillation itself, the actual facts of how bourbon, that hugely popular American take on whiskey, came to be is shrouded in mystery, legend and half-truths. Official records are sparse but after the whiskey rebellion ended in 1794 the industry enjoyed a relatively calm period and it is widely accepted that bourbon whiskey was born along the western frontier during this time in history. The name bourbon of course comes from the French – Bourbon County in Kentucky was named in 1785, in honour of the French royal family and in recognition of the support provided by France during the American revolution. Counties such as Fayette and towns like Paris, Louisville and Versailles were named for the same reason. Bourbon County was initially very large but it was carved up into smaller pieces in 1789, and then in 1792 it became thirty-two distinct counties. However, the entire area was still known as 'Old Bourbon' and whiskey from these areas would often be labelled as such. So, this goes some way to explaining how the name came to be, but not how the style of whiskey that is bourbon came about.

One of the enduring legends is of Elijah Craig, the man who has become known as the father of bourbon. The Heaven Hill Distillery in Kentucky, which today produces the Elijah Craig brand, claims that he was the first person to age his whiskey in charred oak barrels, which gives bourbon its signature taste. This discovery was said to be due to an accident when a fire broke out at the distillery and changed the expression of the whiskey within the scorched barrels. However, there is also an alternative theory, also championed by the distillery,

Elijah Craig, possibly the father of bourbon. *Author's collection*

that Craig was using barrels that had once contained sugar to age his whiskey and he charred the interior of the barrels partly to kill off the old sugar residue but also to strengthen them for use in ageing whiskey. He was impressed with how the finished whiskey tasted, and must have realised he was on to something. Whatever the truth, it is fact that Craig continued to refine the barrel charring process, imparting a smooth rich flavour to the whiskey that would become known as bourbon. These days it is set in law that bourbon whiskey must be aged in new American oak barrels that have been charred by a flame, and it is a great story that the process was originally discovered by chance.

Craig was a Baptist preacher from Virginia with a strong entrepreneurial drive. He would fight to preserve religious freedom in order to ensure that the religious persecution that had occurred in Europe was not repeated in the United States. In 1782, with religious freedom now established in the constitution, Craig moved to Kentucky County, Virginia, where his brother Lewis had settled a year previously. It was here that he set up the Great Crossing Church in 1786 and became chief pastor, but he continued with his business interests alongside his spiritual ones, which garnered condemnation from some within the church. He set up a second church a couple of years later, after being annexed from his position at the Great Crossing. Craig would start Kentucky's first classical school; he donated land to Georgetown College in Kentucky and started the town's first fire service, upon which he served as fire chief.

In 1789, Craig founded a distillery in what was then Fayette County, and it is here that he earned the somewhat dubious title of the father of bourbon. Many whiskey historians maintain that Craig would have been making the same whiskey as his contemporaries, and that the legend of how he became the first man to use charred barrels is simply that, 'a charming legend'. There is no actual historical evidence to suggest that Craig's whiskey was unique at the time, nor that he was using charred barrels. During the period hundreds of small farm distilleries were making corn-based whiskeys which they labelled bourbon to distinguish them from the rye-based whiskeys common in the east.

So, if not Elijah Craig then where did the use of charred barrels to make bourbon come from? One possibility is that cooperages, the companies that made the barrels, were already using fire, water and steam to make the staves pliable so that they could be worked into the right shape for the construction of barrels. It is possible that some of these charred staves simply imparted the desirable flavour to the whiskey and that it was a happy accident. It is, however, fact that French Cognac makers had been using charred barrels for centuries by this point and this knowledge could have come across with the many French who settled in the United States. New Orleans, a major market for the new whiskey, was essentially a French settlement, and given that it was on the river trade route, easily accessible to whiskey entrepreneurs from Kentucky.

Whiskey historians also speculate that the use of previously used barrels may have initially brought about the practice of charring them. Distillers along the western frontier would buy previously used barrels in order to save money but some of these would have previously been used to store fish, pickles, meat, and even soap, so scrubbing them would not have been enough to remove these unwanted residual flavours that would leach into the whiskey during the ageing process. However, by applying a flame to the inside of the barrel the distillers found that not only would this destroy the unwanted flavours but it would also strengthen the barrels and impart more desirable nuances to the finished whiskey. It could have been as simple as that, but whatever the truth there is no escaping the fact that the legend of Elijah Craig is indeed a good one.

We find ourselves on more solid ground with identifying who it was who first coined the term 'bourbon whiskey', at least in marketing terms. In the 1790s Jacob Spears founded the Famous Peacock distillery in Bourbon County, and in a 1821 advertisement for the distillery in the *Western Citizen* newspaper was an advert that listed bourbon whiskey. This is widely accepted to be the first mention of it in print. The belief is that Spears called his whiskey bourbon because he made it in Bourbon County, and this would seem logical, particularly as when Kentucky became a state in 1792 and there was a shuffling of boundaries Spears took to labelling some of his whiskey as Old Bourbon Whiskey. This, of course, refers to the old county lines when Bourbon had been a part of Virginia.

Today Kentucky remains the bourbon capital of the United States, with more than 95% of production from the state, and whether it was Craig or Spears who is the true father of bourbon it matters not, because both had a part to play. Today, with the definition of what is and what isn't bourbon enshrined in federal law, these two men continue to cast a long shadow over the history of the drink.

Let's raise a glass to Spears and Craig.

Chapter 8

Whisky in Pop Culture

He put that bottle to his head and pulled the trigger.
Bill Anderson and Jon Randall

I touched on this subject briefly in the opening chapter of this book, but so widespread has whisky become entwined in popular culture that further study is warranted. The drink has been equally celebrated and vilified in song, prose and poetry. It has found itself the butt of jokes by scores of comedians, inspired novels and appeared in movies as diverse as *High Noon* and *The Shining*.

It is very much a two-way street, though, and switched-on marketing men at whisky companies have not been slow to realise the sales potential of getting their brands into the hands of consumers via mass media. Johnnie Walker bottled several whiskies based around the George R.R. Martin *Game of Thrones* TV series, Sadler's Peaky Blinders Irish Whiskey has become an iconic brand, and a very popular Texas whiskey is the J.R. Ewing Reserve, with the bottle label sporting an image of the South Fork Ranch from the long gone, but fondly remembered, TV soap. Rock stars have even got in on the act, with Bob Dylan taking an active part in the creation of his rather exquisite Heaven's Door range of whiskeys and the powerhouse rock act Metallica have added their name and expertise to Blackened, which is a blend of whiskeys finished in black brandy casks. While this whiskey was being aged, the original distiller, the late Dave Pickerell, blasted the band's music into the room where the casks were being finished. This process is called sonic enhancement and the theory is that the soundwaves cause the spirit to vibrate within the barrel, thus increasing the interaction between spirit and wood. Did this impact on the flavour in any way? Apparently it does, with the low-hertz frequencies causing the whisky within the barrel to reverberate at a rapid rate, interacting with the staves of the barrel and measurably releasing wood compounds and flavours. These claims may sound dubious but it is steeped in scientific fact, and at the very least it's a very good story to tell at your next whisky-tasting session.

> *We wanted to be able to look fans in the eyes, and say this whisky is something we started from the beginning, and it's something that has a Metallica touch to it.*
>
> Metallica's Lars Ulrich

Rock music and whisky these days walk, or should that be stumble, hand in hand and there are brands dedicated to The Rolling Stones, Motorhead and, among many more, the Scorpions. More often than not the bands involved have been a part of the production process from start to finish, though some have admittedly simply allowed their names to be used. You can just imagine Keith Richards standing in the Suntory building and sampling glass after glass before signing off on the finished drink.

> *Scotch and Coke was very much a Beatles drink.*
>
> Brian Epstein

Music and whisky, though, have always had a close relationship but while most whisky songs are about drunkenness and the consequences that follow – cue countless country songs – there are some that rise above this cliché. The late great Leonard Cohen's *Closing Time*, for instance is a beautiful, poetic ballad that perfectly captures the feel of sharing a dram with good company into the early hours. The fiddler fiddles something so sublime, indeed.

However, the relationship between whisky and pop culture is not a new thing, and from the earliest days whisky has held its place in the culture of the times. The ancient English ballad *John Barleycorn*, for instance, tells the story of the cultivation, malting and milling of barley and the enjoyment of the resulting alcoholic drink, but even if it doesn't specify whether the finished drink is beer or malt spirit, the Scottish bard Robert Burns certainly took it to be whisky when he wrote his Scottish version of the ballad. Likewise, the Irish ballad *Whiskey in the Jar* was originally penned around the time of Cromwell's invasion of the Emerald Isle and tells the story of highwayman Patrick Fleming, who was betrayed by his lover, Molly. In reality Fleming was no hero and he robbed the rich and poor alike, as well as maiming many unfortunate souls before being hunted down. However, after being condemned to death he made a legendary escape from his cell by climbing up a chimney and disappearing into the countryside, an event borne out of desperation but forming a legend. Fleming, though, was eventually recaptured and in 1650 he faced the hangman's noose.

In death Fleming has become something of a folk hero, largely because the Irish and Scottish have always liked the idea of these gentlemen of the roads robbing English landlords, and the misty eyes of history have elevated them to the status of true patriots. In the 1970s the Irish band Thin Lizzy created a new version of the song and it remains hugely popular today, being much covered by dozens upon dozens of artists including Brian Adams, The Pogues and Metallica.

> *The true pioneer of civilisation is not the newspaper, nor religion, not the railroad – but whisky.*
>
> Mark Twain

Twain, or to give him his real name Samuel Langhorne Clemens, had a love affair with Scotch whisky that lasted his entire life. He started his drinking career when working as a steamboat pilot on the Mississippi River, but it wasn't until a trip to Europe in 1873 that he first discovered Scotch. He was familiar with bourbon whiskey but as soon as he took his first Scotch, he found that he much preferred the more intense smoky taste that the drink offered. So enamoured with the whisky variety was he that he immediately penned a letter to his wife, Olivia asking her to stock up on the whisky.

> *Livy my darling, I want you to be sure and remember to have in the bathroom a bottle of Scotch whisky, a lemon, some crushed sugar and a bottle of bitters. Ever since I have been in London, I have taken in a wine glass what is called a cock-tail (made with these ingredients) before breakfast, before dinner and before going to bed.*
>
> Mark Twain

The cocktail Twain refers to is the Old Fashioned and from there his passion for the Scottish variety, calling it his pet of all brews, grew, until he became known as a true aficionado of the spirit. He used his trademark wit to provide many memorable quotes about the drink that are oft quoted at whisky gatherings across the world.

> *I always take Scotch whisky at night as a preventative to toothache. I have never had the toothache; and what is more, I never intend to.*
>
> Mark Twain

Twain may have been among the first to provide delightful whisky quotes in a pre-Twitter, or should that be X, age, but he certainly wasn't alone and many writers, actors, broadcasters, musicians – in fact, people from all walks of life – have provided us with memorable quotes about the complex spirit we call whisky.

> *Happiness is having a rare steak, a bottle of whisky and a dog to eat the steak.*
>
> Johnny Carson

> *I like my whisky old and my women young.*
>
> Errol Flynn

A good drop of hot whisky before bed. It's not very scientific but it helps.

Alexander Fleming

Whisky is liquid sunshine.

George Bernard Shaw

I think I'll have another before tackling the next chapter.

Gary Dobbs, the author of this book

The light music of whisky falling into a glass – an agreeable interlude.

James Joyce

And, the whisky quote to end all whisky quotes comes from Humphrey Bogart, whom, when on his deathbed, is reported to have said, 'I should never have switched from Scotch to Martinis.'

Whisky and the movies have always gone hand in hand. From the earliest celluloid westerns where cowboys slugged back the drink, served in a dirty glass, to clear away the trail dust, to the hard-boiled detectives moodily sipping from an ice-clinking crystal glass, whisky has held its place; if not in the cast list, then at least in the props department. A glass of whisky in the hand is often used as a kind of visual shorthand to suggest the attributes of a character – powerful, sophisticated, important, ruthless, even evil. From the dizzy heights of Hollywood to the urban soundstages, the water of life has always played a pivotal part.

Though far from a bit player, there have been several movies that have used whisky as the fundamental core of the plot. Perhaps the best known is *Whisky Galore* (1949), which was based on the novel of the same name by Compton Mackenzie. The novel that served as the source material was in itself based on an actual event that occurred in 1941 when the SS *Politician* ran aground on the island of Eriskay, in the Western Isles of Scotland, with a cargo that included 28,000 cases of malt whisky. Official records show that the ship, as well as being laden with the water of life, was carrying 290,000 10 shilling notes, which today equivalates to several million pounds.

The man at the helm was Captain Beaconsfield Worthington and when the ship came upon a ferocious storm, he realised that he would have to change course if he was to avert disaster. This presented yet another danger to a crew who were already anticipating having to cross a U-Boat-infested Atlantic, but things would get worse with the ship blown even further off course until it ran aground on the sand banks at Rosinish Point off Eriskay. With the fuel tanks ruptured and the engines stalled, the crew could do nothing but await rescue.

The entire crew were eventually rescued but this was not without drama and because the ship had gone so far off course rescuers initially went to the wrong location after receiving its distress call. Thankfully an islander had noticed the ship starting to list and quickly raised the alarm.

As soon as news of the ship's cargo spread there were many unofficial salvage operations, with locals collecting as much of the whisky as they could before the authorities arrived and closed down the shenanigans. To many, this was not seen as stealing; as far as they were concerned once the goods had entered the sea, they were ripe for the taking under old maritime salvage laws. However, the authorities, those stern men from the customs and excise, did not see it this way and many homes were raided on the isle by customs men and police in search of the missing whisky. In all it was estimated that

The ill-fated SS *Politician. News Chronicle*

local people had taken more than 24,000 bottles of whisky from the wreck before the hull was dynamited to remove temptation.

> '*Dynamiting whisky*,' islander Angus John Campell commented to the *News Chronicle*. '*You wouldn't think there'd be men in the world so crazy as that.*'

Was all the whisky destroyed?

It would seem not and as recently as 1987, a South Uist man, Donald MacPhee, found eight bottles that the sea had given up after yet another storm. The bottles were sold at auction for a respectable £4,000.

Today the wreck of the *Politician* remains off the coast of Eriskay. Its deck and cabins have long been destroyed by the relentless sea, but its name lives on with the Am Politician pub and restaurant, which does a roaring trade thanks to the legend of the ship that still brings tourists to the area.

Both the 1949 movie classic and the rather pointless 2016 remake take a comedic approach to the story, with the locals defying the rather pompous excise men as they come up with all manner of outrageous plans to save the precious whisky from a watery grave.

> *Northwest of Scotland on the broad expanse of the Atlantic lie the lovely islands of the Outer Hebrides. Small, scattered patches of sand and rock rising out of the ocean. To the west there is nothing except America …*
>
> *Whisky Galore*, 1949

The 1949 movie may have had a first-time director at the helm with a minimal budget, but it took everyone including the studio by surprise and remains Ealing's most financially successful release, and arguably one of their best known and most entertaining. It was also the movie that saw Ealing make waves at the US box office with the film, despite the bizarre title change to *Tight Little Island*, being a massive success. The film played well in all international markets, being hugely successful in France where it went by the title *Whisky a Gogo*.

Whisky Galore original cinema poster. *Ealing Studios*

It's a light-hearted movie certainly, but the performances are universally excellent, the script faultless and the black and white photography particularly atmospheric. The viewer finds themself cheering with the villagers as they manage to outwit the excise officers and when the credits roll the audience are left with a smile upon their faces, as well as a thirst that only a whisky can satisfy.

No discussion of whisky in the movies would be complete without mentioning Ken Loach's excellent 2012 *Angel's Share*. The film is basically a heist comedy, but with Loach's eye for social realism and writer Paul Laverty's understanding of character it is a minor masterpiece. Don't just take my word for it; it took the jury prize at the 2012 Cannes Film Festival.

The main role was taken by newcomer Paul Brannigan and it would turn out to be inspired casting. The young, inexperienced actor brought the right amount of grit to the role of Robbie, a young man who has escaped prison by the skin of his teeth and finds himself having to serve out 300 hours of community service. It is while carrying out his service that Robbie meets up with several other misfits and these form the core cast of the movie.

It is only when the group are taken on a distillery tour by their kindly supervisor that it is discovered that Robbie has an incredible talent for nosing whisky.

The plot of the film saw a cask of the fabled Malt Mill whisky discovered in a warehouse and this is tipped to fetch a record price at auction. The plan, dreamt up by our loveable losers, is to break into the warehouse prior to the auction and steal an indiscernible amount of whisky from the cask – just one bottle will bring them a life-changing amount of money and nobody would be the wiser; all they would be taking is the angel's share. The perfect crime.

The imaginary cask of Malt Mill used in the film was based on an actual whisky that was believed to have been lost to history. Malt Mill came about in 1908 following a falling out between the owners of the Lagavulin and Laphroaig distilleries. In 1907, following the death of Laphroaig owner Alex Hunter, his nephew Ian Hunter was

The troubled young man Robbie discovers he has a rare nose for whisky in Ken Loach's *Angel's Share. Joss Barratt and Sixteen Films*

brought in to continue in the family tradition. However, a dispute soon broke out because Hunter felt that Laphroaig was not getting the best returns because it was being overshadowed by Lagavulin and the two companies drifted apart following a complicated legal battle.

Peter Mackie, owner of Mackie and Co, had been using both Lagavulin and Laphroaig in its popular blends, notably White Horse and Mackie's Ancient Scotch, and so enraged was he by the legal battles that he started his own distillery in the grounds of Lagavulin and set about creating a replica of the Laphroaig whisky. He called his distillery Malt Mill and the whisky produced was used in his popular blends, and would continue to be used until the distillery closed in 1962, with its equipment and premises merged into Lagavulin. None of the whisky produced at Malt Mill was ever bottled as a single malt and until the success of the *Angel's Share* movie nobody, outside of a select bunch of whisky experts, had ever heard of Malt Mill.

However, they say life imitates art and that is exactly what happened following the release of the movie, when a bottle of Malt Mill suddenly appeared on the whisky scene. The miniature bottle, which would have been sent out as a taster for blending houses, sold at auction for £6,670 – that's 5cl, a double measure of whisky. A pretty expensive drink, and it is beyond doubt that its worth only came about because of the success of the *Angel's Share* movie.

What better recommendation could there be!

Mackie's Ancient Scotch.
Author's collection

Chapter 9

Prohibition and the Rise of the American Gangster

It is teaching the public to drink secretly, it is causing gambling and immorality in the homes, it is bringing blindness to scores, it is causing those who used to take a nightcap before bed to take dope, it has caused a large increase in the sale of narcotic drugs and daily increases in the army of dope fiends. It is making imbeciles of many and before long the asylums for the insane will be overfilled. It is causing diseases of the liver, kidneys, eyes and heart – all due to the consumption of home brews containing fusel oil, wood alcohol, copper and other deleterious substances.

Are we physicians going to make a stand for public health and for our country, or are we going to sit back and suck our thumbs? We must decide whether prohibition and home brew, or the use of light wines and beer properly made under government standards is the greater evil.

Dr W. Wallace Fritz, speaking in June 1921 on the effects of prohibition in the United States

Prohibition came about in the United States after more than a century during which the strength and reach of the temperance movement was felt across the country, from the fast-growing cities to the remotest rural hinterland. The religious revival that spawned the movement had been gaining support across the new nation, with its strict codes of morality and self-conduct taking root among communities and spreading with the vivacity of a righteous pandemic. The organisation was initially most influential in rural areas, but soon the messages against drunkenness and immorality began to gather significant support in larger areas and this eventually prejudiced the thinking of elected officials, who recognised the voting power of the temperance factions.

This new-found political support empowered the movement and soon the message of moderation became one of total abstinence, a message carried from town to town by fervent preachers and activists. What had started out as a noble cause that didn't prohibit all alcohol – only the hard stuff such as rum and whiskey were out of bounds while light beers were seen as acceptable – soon became one of total avoidance. Alcohol was fast becoming the demon drink, responsible for poverty, violence, idleness and general degradation.

Cartoon by the Anti-Saloon League. *Westville Public Library*

Although national prohibition would not come into effect in America until 1920, the first prohibition law at state level was passed in Maine in 1851. In Portland alone, then home to fewer than 10,000 people, there were 200 licensed liquor stores and a good few hundred that were unlicensed. This victory, following a period of intense campaigning against the perceived evils of alcohol, was a turning point for the temperance movement and strengthened its position. Campaigning had resulted in a state outlawing the sale of intoxicating beverages for the first time anywhere in the United States, which meant that the temperance movement had gone from being seen as an organisation of cranks to a powerful force for change.

In 1855, four years after the law had passed in Maine, there was a riot on the streets of Portland in which the state militia were ordered to fire on an angry crowd who had gathered outside City Hall to protest against prohibition. One man was killed and several more injured, but still the mayor, Neal Dow, often called the father of prohibition, remained steadfast in the face of the criticism of what had by now become known as the Maine Law. He himself was a lifelong teetotaller and had been one of the co-founders of the Maine Temperance Society.

The riot, now known to history as the Portland Rum Riot, had all kicked off when a group of working-class Irish residents heard the rumour that Neal Dow, the teetotal mayor, had stashed a large quantity of alcohol at city hall for the consumption of visiting dignitaries.

'Hypocrite,' the gathered Irish crowd shouted out and demanded the mayor be arrested, after all it was he who had championed the law that prohibited the sale and manufacture of alcohol in the state.

The crowd quickly grew to more than 3,000 and rocks were thrown at the building as they demanded to be allowed to search City Hall for the booze. The leader of the protesters showed a warrant issued by a local judge that gave them a legal right to search City Hall, but this was refused, which only served to anger the crowd further. It was a curious law at the time, but if more than three men believed a crime had been committed, they could

Mayor Neal Dow, often called the father of prohibition. *Author's collection*

approach a judge and obtain a search warrant. The men behind the riot had done so and received such a warrant but the police at City Hall dismissed it out of hand. Things began to escalate, and in fear of the crowd rushing the building, Dow called out the armed militia and shots were fired. One man, a John Robbins who was due to be married the next day, was killed. The rioters quickly dispersed, but the events of that day ruined Dow's political aspirations; he lost re-election by a wide margin and a year later the state repealed the prohibition laws.

A few years later the temperance movement found its cause overshadowed by the fight to abolish slavery and the civil war that was fought to settle the matter. In 1862 the federal government, hungry for funds to bankroll the war, legitimised the liquor trade by introducing a series of licence fees as well as a taxation on the production of intoxicating beverages. Over the next few years almost a third of the federal budget came from alcohol taxation, and temperance was brushed aside as the country faced the horrors of a civil war. The earlier whiskey rebellion had almost brought the country to war with itself, but this time it was all out conflict between the North and South, with America seemingly tearing itself apart.

Following the war, the cause for temperance was once again revived and the strength of organisations such as the temperance movement and the Anti-Saloon League would continue to grow. Demonstrations, with women becoming human barricades outside saloon doors, became common place. Often violence would be committed against these women by both saloon owners and patrons, who became enraged when they found their entry blocked. These women were spat upon, kicked, punched and dragged aside but they remained committed to their cause.

> *Terrible saloons may be, but that just meant that they were terribly wonderful.*
>
> Author Jack London

Carrie Nation, also known as Hatchet Granny and a member of the temperance movement, soon realised that peaceful marches and demonstration were all very well, but if change was truly desired then a more direct approach was required.

Carrie Amelia Moore was born in 1846 in Garrard, Kentucky, to a wealthy slave-owning family on a large farm, but the civil war brought ruination to her family and when she was 21, she moved to Missouri. There she married a young doctor and veteran of the civil war named Charles Gloyd. At first theirs was a happy marriage but Gloyd, who had developed a taste for whiskey during his time in the military, soon found himself unable to control his drinking, and his life began to unravel. It was not long before his work suffered and in response to this he simply drank even more and allowed life to pass by in a whiskey-infused haze. He was by now struggling to support his pregnant wife, and there were many arguments between the couple. Carrie, though, was made of sterner stuff than her husband and eventually she left him. She went to live with her parents, who had also settled in Missouri, and it was here that she gave birth to a daughter, whom she named Charlien. If Carrie had hoped that naming the baby after her estranged husband would bring him to his senses, then her hopes would soon be dashed. Charles Gloyd died, penniless and drunken, less than a year after his daughter had been born.

Carrie, despite initial depression at the loss of her husband, was able to rebuild her life, and she soon found employment as a teacher so that she was financially able to care for both herself and her daughter. She would later marry a lawyer and preacher named David Nation. The young couple moved to Texas, where they took over a cotton plantation, but when the business failed they packed up and moved to Kansas. This time the marriage was successful for a number of years, but the couple would divorce in 1901 after Carrie started her infamous crusade of smashing up saloons with a hatchet. Carrie became famous for her crusade against drinking establishments but her husband was less than impressed and he sued for divorce on the grounds of desertion when his wife was away on another of her saloon-destroying rampages. David's death came in 1903 from complications surrounding a stomach ulcer, but by now his former wife was gaining national fame as a radical member of the temperance movement. The report in the 7 October edition of *The Barber County Index* reported that Nation's doctor had been unable to do anything other than provide pain relief during the man's final few hours. The report then went on to cover the antics of his now infamous former wife.

Carrie Nation's rise to become the scourge of saloon owners had started in 1900 when, claiming she was acting on a divine message from above, she went to Kiowa in Kansas, entered a saloon and told the owner, a Mr Dobson, to stand aside as she was going to smash the place to pieces. She was carrying several rocks wrapped in paper and she started to throw these at the saloon's fixtures, causing considerable damage. Once that was done, she calmly walked out and found another saloon, and with yet more rocks she repeated the process. All in all, she attacked three saloons on that June afternoon in 1900.

Afterwards, she baited the sheriff to arrest her but the lawman would not act, and then she moved on to Wichita, but this time she carried a hatchet with her instead of a pile of rocks. She was arrested several times, but all this achieved was to spread her fame, as well as earning her the nickname Hatchet Granny. Soon a band of supporters joined her and would stand outside saloons singing hymns while Carrie took her hatchet to the fixtures and fittings.

Carrie Nation was jailed and fined many times, but this never slowed her down. Her fines would be paid by supporters or by revenue collected from the sale of stick pins in the shape of hatchets that bore the words 'Death to whiskey'. She also, during this period, received a reliable income from the lecture circuit, even appearing at rallies in Europe, during which she made a strong case about the evils of alcohol. Her fame continued to spread across the United States and the world beyond, and she would continue her crusading until she died in 1911.

In 1917 the political influence of the various temperance societies, especially the Anti-Saloon League who by this time had become a powerful political force, was felt when the Eighteenth Amendment passed both chambers of Congress. This in itself was incredible since the entire constitution was intended to extend and safeguard the freedoms of citizens, and yet this new amendment was designed to severely limit certain freedoms. Initially this was seen as a wartime measure to preserve grain, but the skilled voices of the prohibitionists – particularly Wayne Wheeler of the Anti-Saloon League – were able

The Hatchet Granny. *Author's collection*

to exploit the considerable anti-German-American sentiment against the beer makers. It was clear that wartime prohibition was not enough to satisfy the prohibitionists. The amendment was therefore later ratified in 1919 and came into effect a year later on 17 January 1920.

Effectively the alcohol industry, the country's fifth largest revenue provider, was now dead.

The Eighteenth Amendment, known as the Voldstead Act after its promotor Andrew J. Volstead, limited the manufacture, sale or transportation of intoxicating spirits, which resulted in the almost total shutting down of the spirits industry. Distilleries across the country had to shutter their doors and lay off their workforce, and in fact the greatest majority of these distilleries, some with a rich history that stretched back to the colonial period, would never return. Indeed, even today the effects of prohibition, that noble though failed experiment, are still evident in the names of the long-vanished distillers and brewers.

Prohibition, though, was absolute but did allow several licensed distilleries to continue operating in order to produce medicinal whiskey, which could only be obtained with a doctor's prescription. This medicinal whiskey could be prescribed for any number of ailments, including blood pressure, common colds, headaches and even stomach upset, and pretty soon it became evident that doctors and pharmacists were raking in huge amounts of cash by serving their thirsty patients, or perhaps customers would be more apropos given the high levels of corruption and faked ailments involved. One household name that benefited from prohibition was Walgreens. The pharmacy chain had been started in 1901 and at the time of the Volstead Act operated close to thirty stores, but by the end of the first decade of prohibition it boasted around 400. Today Walgreens is the second largest pharmacy chain in the United States, with its international operations including Boots, a name that graces most high streets across the United Kingdom. The company's greatest period of growth remains the 1920s, when it benefitted from selling medicinal spirits, mostly whiskey, at vastly inflated prices and its annual sales at the end of the decade were around $4 million.

It wasn't only the pharmacists and doctors who cashed in on prohibition, though, and soon a new kind of criminal was operating. The bootlegger became a folk hero to many as a vibrant illegal economy was born with the aim of giving the public what they wanted; something that was being denied to them by an increasingly authoritative federal government. Organised crime, while always present in the big cities, now found a firm foothold right across America and its tentacles soon spread from coast to coast.

Before prohibition organised crime, while present, had been far from organised, with gangs of varied ethnic origins, Irish, Italian, Jewish and Polish, all acting independently of each other and the bulk of their activities in loan sharking, protection rackets, robbery, extortion and contracted violence. Prohibition, though, saw gangs becoming far more coordinated, working together to shift smuggled alcohol through each other's territories and into the hands of eager consumers across the nation. It is a fact that terms such as syndicates and organised crime did not come into popular usage until after prohibition had started. The criminals simply had to become organised and they all had to share the business costs such as bribing politicians, policemen, judges, juries, witnesses and even federal prohibition officers.

Revenue from smuggling and selling illegal spirits were massive and soon the criminals were hiring accountants, lawyers, truckers,

Al 'Scarface' Capone, arguably the best-known gangster of the era. *Getty Images*

boat captains and warehouse men, and an army of violent thugs who were sent out to intimidate, threaten or kill anyone who stood in the way of the ever-growing illegal booze trade.

The earliest bootleggers smuggled foreign-made commercial spirits into the United States across both the Canadian and Mexican borders and then along the sea coasts in ships that held a foreign registry. The ships would often stop offshore, just outside of the 3-mile limit where the government lacked jurisdiction, and then transport their illicit goods to shore in high-powered craft that could easily outrun the coastguard. However, when the coast guard started searching ships outside of the 3-mile limit and using high-powered craft of their own the bootleggers found they had to find alternative ways to meet the great demand for spirits. The vast sums of money involved ensured that the bootleggers would find a way, no matter what obstacles were thrown in their way, to get their spirits into the hands of the consumer. Woe betide the foolish prohibitions officer who tried to confront the heavily armed and well-organised smugglers. Boats were run out far into the ocean to buy booze from Great Britain and Canada, which led to the term 'rum running', and people were paid by the mobs to set up small stills that churned out foul-tasting spirits in their kitchens.

Illicit drinking establishments, known as speakeasies, had sprung up right across America, and it is estimated that by the end of the decade there were more than 30,000 in New York alone. These ranged from dingy basements to opulent clubs with ballrooms and resident jazz bands, and as these venues exploded in popularity then so too did the influence of organised crime. The mobster, the gangster, became an image of sophistication, a lovable rogue who outwitted the authorities in order to bring a product to a thirsty populace. However, this was far from the reality of the situation and there was nothing romantic about the mobsters who quickly took control of the illicit spirits business. In fact, within minutes of the amendment becoming law a truck carrying medicinal whiskey was hijacked in Chicago by an armed gang, and that same day a government-bonded warehouse containing grain whiskey was broken into and emptied in a daring raid.

Before looking at the way crime organised itself in order to exploit the new laws, it is worth noting that prohibition was not limited to the United States. Iceland went dry from 1912 to 1932, Finland from 1919 to 1932 and Russia, ignoring the national love for vodka, brought in prohibition between 1914 and 1925. Australia and New Zealand also came close to total prohibition but failed to secure enough votes to push the laws through. It is also worth noting that by the time America finally went dry, twenty-one states had already done so. The consequences of prohibition would have hardly been considered, nor in fairness could they have been foreseen, at least not to the extent of what occurred during those thirteen years that the law stood. The signs were there though. American lawmakers seemed to have been remarkably insular during the period, and chose to overlook the fact that when Tsar Nicholas II closed down Russia's vodka distilleries, production went underground and resulted in the state losing a third of its income from tax revenue.

American prohibition did not outlaw the purchase or consumption of alcohol, but instead the Eighteenth Amendment banned its manufacture, distribution and sale, though as noted even this was not absolute and the limited production of medicinal alcohol was allowed to continue. The warehouses where this spirit was stored soon became a target for mobsters.

George Remus, a defence lawyer from Chicago, soon saw a way of making a fortune due to the new prohibition laws. When he realised the vast sums of money the bootleggers he defended were turning over, he moved to Cincinnati and, using a loophole in the Volstead Act, started to purchase both distilleries and small pharmaceutical companies. He would use these companies to obtain permits to withdraw medicinal whiskey from the warehouses of his own distilleries and then arrange for the whiskey to go missing. His men would load up the truck at the warehouse and later be robbed on route by even more of his men. He now had a supply of whiskey to supply to his many buyers. No sooner had the barrels been reported stolen than the cycle would start all over again. So lucrative was his operation that at one time the defence lawyer had 3,000 men working for him either on the black market or in one of his 10 distilleries. He was soon turning over millions of dollars each year and ended up owning many of America's most famed distilleries. With the great wealth, personal fame soon followed and author F. Scott Fitzgerald loosely based aspects of his novel *The Great Gatsby* on Remus's life.

> *The more I studied the Volstead Act the more convinced I became of its frailties and I decided to get in on the ground floor.*
>
> George Remus

For a period, everything seemed to go his way, but eventually Remus and his flamboyant lifestyle came to the eye of the authorities. In 1925 he was arrested for numerous violations of the Volstead Act and sentenced to two years in a federal prison. While serving his sentence he befriended another inmate, and so close did the two become that Remus confided in his friend that he had managed to hide much of his money, which was now controlled by his wife, Imogene. However, the inmate who had gained Remus's confidence was actually an undercover prohibition agent who had befriended him in order to find out information such as this. However, the agent, Franklin Dodge, did not pass this information on to his superiors but instead resigned his post and began an affair with Remus's wife.

Franklin and Imogene quickly liquidated Remus's assets, even selling the famed Fleischmann Distillery, and hid as much of the money away as possible. While Remus served the rest of his sentence the pair hired a hit man for a fee of $15,000, an insignificant amount of money given the vast fortune they had siphoned from the Remus estate. However, the assassin, fearing a double cross, decided at the eleventh hour not to go through with his task and instead he told George Remus of the plot against him. When Remus was released from prison, he was furious and swore vengeance on both his wife and the one-time prohibition agent.

In October 1926 Remus was on his way to court in order to finalise his divorce when he spotted a cab carrying his estranged wife. He ordered his driver to follow the vehicle and force it to pull over at the side of the road. Remus immediately jumped out of his own car and walked over to the cab, pulled a pistol from his pocket and fatally shot Imogen in the stomach. Incredibly, Remus avoided jail for this cold-blooded murder by defending himself and claiming transitory insanity, a defence he had used successfully in the past while defending several of his own clients. He was sent instead to an asylum but was released after only seven months when it was decided that his mental state was greatly improved.

Perhaps the most famed of the prohibition-era mobsters was one Alphonse Gabriel Capone, otherwise known as Scarface on account of a wound he had received during a knife fight with the brother of a woman he had insulted. Capone hated the nickname but even he had to admit that it gave him an air of danger, which proved useful in his criminal career. This career saw him running a multi-million-dollar empire that controlled prostitution, gambling and bootlegging in Chicago but whose appendages spread out across the United States and beyond. In fact, during the height of prohibition Capone is known to have travelled to Scotland under an assumed name to meet with associates who kept a steady flow of Scotch finding its way across the water to Capone's operation. While there he indulged in his favourite game of golf, playing at several of Scotland's best courses including the Old Course at St Andrews.

Capone was born in 1899 in New York to Italian immigrant parents. His father, Gabriele, was a barber and his mother, Teresina, was a seamstress. Both parents were hard-working, honest people and young Alphonse was their fourth child. In total the couple had nine children and did their best, providing a comfortable if frugal life for their brood.

During his early years the young Capone was a perfectly ordinary child, and by all reports a diligent scholar, but by his early teenage years he had started to rebel against a school system that felt far too authoritative for his liking. He started to miss school, spending his days with the street gangs indulging in acts of vandalism and petty crime, and the young Capone soon gained a reputation for being a tough guy after several fights with rival street gangs. At the age of 14 Capone was expelled from school after assaulting a female teacher who had tried to punish the young boy by caning him across the hand with a ruler. Capone endured several of the swipes from the ruler before deciding enough was enough and punching the teacher in the face. 'She went down like a sack of shit,' Capone later boasted.

From that day on, the young Capone drifted into the world of crime, at 16 he was a member of Manhattan's Five Points Gang and served as both a bouncer and bartender at the Harvard Inn, a cheap bar and brothel on Coney Island. The bar was owned by mobster Frankie Yale and, seeing promise in the young Capone, he took him under his wing. It was during the period that Capone was slashed across the face by Frankie Galluccio because Capone had insulted his sister. Reports from the time state that Capone leaned across a bar and told the woman, 'You've got a nice ass, honey,' before her brother went berserk, produced a knife and gave Capone the marks that would earn him the despised nickname Scarface. Given Capone's later reputation it is surprising that he never sought revenge against Galluccio and simply shrugged off the attack. Maybe his own sense of mobster morality told him he had been in the wrong for his comments to the lady and he simply chalked it up to experience.

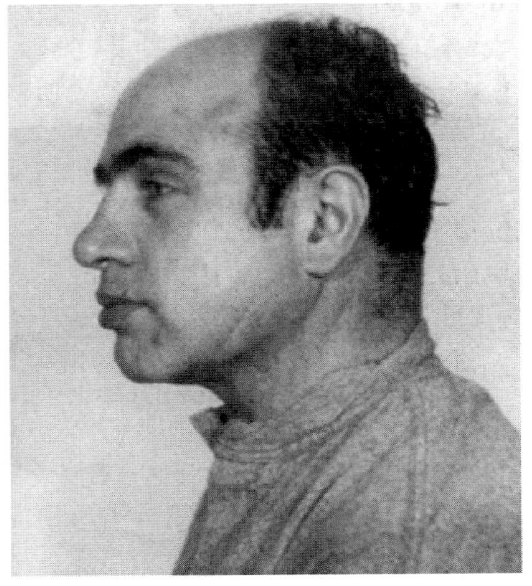

Those infamous scars that gave Capone his despised nickname. *Author's collection*

The mob theory was just that – Capone had violated

mob ethics by being so familiar with Galluccio's sister, and seeking revenge for his scars would likely bring about repercussions. Capone wisely let the matter go, but would forever carry a reminder of that day upon his face. Galluccio was worried, particularly when Capone became a big name in the underworld, but in a press interview with the *Chicago Evening Tribune* many years later he said of the incident, 'Fuck him, he deserved it.'

By 1919 Capone was living in Chicago with his wife, Josephine, and it was here that he started to build a solid reputation for himself. He was thoroughly brutal in the way he took care of business and soon became an enforcer for crime boss Jim Colosimo. This led to Capone being reunited with Johnny Torrio, a small-time crook from his days in New York. In 1920 due to a power struggle between rival mobs, Colosimo was shot down. Capone was rumoured to have been the man on the trigger, but whatever the truth Capone and Torrio took over the operations and the gang became known as the Chicago Outfit. Now Capone and Torrio ran most of the bootlegging operations in the south side of the city, and in the process built up an empire that brought in $70 million a year. Together, the men ran brothels, speakeasies, illegal gambling and, of course, bootlegging, and the two were wise enough to allow other gangs to share in the spoils in order to avoid bloodshed, which was bad for business and brought unwanted attention. However, the gangs that ran with the Outfit were mostly Italian-Americans, and trouble did often flair up with the Polish and Irish groups who controlled the north side of the city. Between 1920 and 1926 mobsters killed 315 of their own and the police were themselves responsible for the killing of another 160.

In 1925 Torrio was shot several times when rival mobsters attacked him and although he eventually recovered, he decided enough was enough and handed over total control of the lucrative operations to Capone. He moved back to Italy, where he resided for a few years before returning to New York in 1928. He died in 1957 of a heart attack while having his hair cut.

Capone enjoyed the high life and courted publicity, indulging himself with the fineries of life including custom suits, expensive cigars, fine food and drink. He would always make himself available to newspaper men who wanted to talk about his success.

'I am simply a businessman giving the people want they want,' he would boast to journalists. Invariably this would be followed with him posing for a photograph. This made him unusually visible for a mobster, a celebrity equal to any of the Hollywood stars who filled the silver screen and quite possibly one of the richest men in America.

Capone now ruled over an empire that took in gambling, prostitution, bootlegging, narcotics, murder and protection rackets, and it seemed that the law couldn't touch him. His public reputation grew and, thanks to a lot of charity work, many began to see him as a Robin Hood-type figure, and the press were not slow to promote the illusion. Huge sums of money came his way and he was raking in as much as $100 million a year; in 1926 it was estimated that Capone was paying out $500,000 a year in bribes to police and city officials to turn a blind eye to his illegal operations.

The public perception of Capone and mobsters in general would change following the violent events of 14 February 1929, when seven members of the Bugs Moran gang were gunned down in Chicago's north side. Capone had been at his home in Florida at the time of the shootings, but there is no doubt that he was then man who gave the order to hit the rival gang. Still, the law enforcement agencies could not link Capone to the brutal events of that cold winter morning.

Despite all the bootlegging, all the murders and the crimes without number, it would be a charge of income tax evasion that finally brought Capone down. He served out most of his sentence in the formidable Alcatraz, and by the time he was paroled in 1939 he was a broken man. Capone had been diagnosed with syphilis during his first few months of imprisonment and by the time he was released the disease had already gone to his brain. The one-time mob boss would die in his bed in 1947 of a heart attack, and medical reports stated that at the time Capone had had the mentality of a 12-year-old. His funeral would be one of the biggest that Chicago had even known, and to this day whenever the name of the city is mentioned Capone springs to mind.

The connection between prohibition and the strengthening of organised crime may seem obvious, but at the time the mere suggestion of cause and effect was dismissed out of hand. And although there were many who felt the prohibition laws had failed, there was no real weakening of resolve on the issue until the market crash brought about by the Great Depression. The financial strength of the country was weakening by the hour and politicians started to consider the vast tax dollars the alcohol industry could bring to a stagnant economy. In 1932 Franklin D. Roosevelt became the thirty-second president of the United States, having run on a campaign that promised to repeal prohibition.

In 1933 he did just that, and for the first time in thirteen years Americans could legally purchase a drink, not that most of them had ever stopped drinking and throughout the years of prohibition the bootleggers had kept a drink in reach of the American people.

Chapter 10

The Wide World of Whisky

Not that long ago there was Scotch, with everything else coming in second but these days that's not the case and whisky, whiskey even, is truly global. Not only are there high-quality American and Irish versions of the drink that are able to compete with the ubiquitous Scotch, there are also decent, often superb, malts, blends and grain spirits coming from Canada, Japan, Wales, England, India, Germany, France, Mexico and even further afield. The ancient art of distilling and ageing fine whisky is now a worldwide endeavour.

Scotland is, of course, famous for its high-quality blended whisky but it is truly the single malt style that distinguishes the country. Blends may outsell single malts but the single malt is as synonymous with Scotland as the tartan that adorns the white-bearded bagpiper on the bottle label, the storm-lashed distillery on the coast line, or the majestic stag peering through the early-morning mist of the Highlands. Single malts may only account for a small, though ever-growing, percentage of the overall whisky business but they do snap up almost all of the kudos and conversation. It is truly an absurd situation and the idea of blends versus malts is quite ridiculous given that the same master blender who spends one day vatting (basically blending) up casks of expensive single malt will also be responsible for blending the more affordable mixtures that actually finance the industry. However, the public perception is, and this is often encouraged by the marketing men, that single malts are more expensive than blends because they are better.

Where Irish whiskey uses both malted and unmalted barley and is largely distinguished by triple distilling, and American

Whisky is now produced across the world. *Author's collection*

bourbon uses a combination of corn, rye and wheat in their mash bills, Scotch single malt is a more singular concoction. It is the product of just one distillery, with the resulting spirit being aged in oak barrels for a bare minimum of three years, though the more premium expressions spend a lot longer in the wood. Single malts, though, are still in a very real sense a blended whisky, unless they are labelled single cask, of course, with several casks blended together to produce the final malt. Each cask will have come from the same distillery, which is what makes it a single malt, but it's still a blend.

Oh yes, it's still a blend.

However, Scotland does not hold a stranglehold on the single malt style – there can be no copyright on a style, it can be produced anywhere and this is increasingly the case in the world of modern whisky. In fact, malt whisky is now produced in several dozen countries, and even Scotland's closest neighbours, the English, are now producing single malts of their own.

Japan was the first country to come to prominence with its own take on the single malt style. The country's first single malt distillery, Yamazaki, having opened in 1923, celebrated its 100th birthday in 2023. The location of the distillery was very carefully chosen and built in an area of Japan that most resembled the Scottish Highlands in both topography and climate, and this seemed to work because the distillery was soon releasing whiskies that could stand proudly alongside the Scottish variety. The company used skills learned from the Scottish, and closely followed their process to produce their whisky, as did the other producers who were not slow in following their lead and now with barely a century of whisky history the Japanese are constantly bringing out expressions that win many of the major international awards for excellence. It seems that, despite the relative youth of Japan's whisky industry, it is still capable of producing spirits that are not just good but quite excellent. In fact, some Japanese expressions are considered among the best in the world. That's quite an achievement given that the whisky they are emulating boasts a history that stretches back far into the distant past.

Back in 2005, close to twenty years ago now, I first tasted Japanese whisky when a friend of mine brought back a bottle from a whisky fest he had attended in Chicago. The whisky in question was a Suntory Yamazaki, and I had no idea what to expect. I was likely quite dismissive, when I first raised that glass to the nose and took in the rich aromas, but when I dared take a sip, I was literally floored. Was this really a whisky from Japan? It was quite incredible, smooth with its flavour components in perfect balance. In short, it was excellent stuff and from that day to this I remain a fan of Yamazaki and pick up a bottle on the all too rare occasions when I can afford it, or I can actually find a bottle. Demand has long outstripped supply, for while distillation can be stepped up to meet a growing demand, ageing the spirit takes time and can't be rushed. This has resulted in many Japanese expressions becoming much more pricey and harder to find.

Although the Japanese looked to the Scots for inspiration in their whisky production, even spelling their product without the 'e' as is the Scottish way, it was the Americans who first introduced the drink to the land of the rising sun.

From 1633 Japan had operated under the Sakoku policy, which severely limited any interaction with most other countries. It was an isolationist ideology designed to protect Japan and banned commercial and private trade with the West, as well as forbidding citizens from travelling abroad for trade it also prohibited nearly all foreign nationals from entering the country. This state of affairs would remain for 200 years until in July 1853, Commodore Matthew Perry led American warships into Edo Bay and demanded he be allowed to deliver a letter to the shogunate.

Perry had his ships train their cannon on the town of Uraga, and as a show of force he ordered his men to fire blank shots. This was gunboat diplomacy, indeed.

These events had been building up for many years and with increasing commerce between American and China, which saw the

An excellent though inexpensive expression from Suntory. *Author's collection*

presence of American whalers in the waters off Japan, it was felt by the American government, still somewhat guided by the concept of Manifest Destiny, that it was their duty to impose Western values on people who were considered primitive Asians.

The Japanese policy of isolation, though, had already been under threat for many years, and as far back as 1844 the Dutch King William II had urged the Japanese to cease their isolationist thinking before they were forced to do so by outside hands. Indeed, in the years leading up to the Perry expedition numerous American ships had been turned back by the Japanese, who remained steadfast in the belief that allowing outside trade would be detrimental to their way of life and tear apart their very culture.

This time the Americans would not be turned back and Perry made it clear that, if necessary, he would use lethal force in order to get his letter from the American president into the hands of Japan's rulers. Eventually, after much debate Perry was allowed to come ashore and meet with the shogun's advisors. He went with considerable pomp, almost 300 sailors accompanying him, and after a thirteen-gun salute he handed over his letter to the Japanese delegates. He told them he would return one year later to receive the answer to the invitation to open trade with the West.

Perry returned early when in February of 1854 he turned up with even more ships and far greater firepower. He was eventually allowed to land on Kanagawa and on 13 March a convention was signed that ended more than 200 years of Japanese isolation. When a delighted Perry left, he gave the Japanese many gifts including a 110-gallon barrel of American whiskey.

The Japanese remained unsure of the benefits of trading with foreign powers but they did at least like the whiskey. Liked it very much indeed, so much so that they were soon trying to make their own version of the spirit by mixing alcohol with spices, sugars and other flavourings but this was far from authentic whisky.

In 1918 chemistry student Masataka Taketsuru went to Scotland to learn the art of distilling. While there Taketsuru enrolled on a chemistry course at Glasgow university but soon took time off from his studies to act as an apprentice at Speyside's Longmorn distillery, where the young man diligently made notes of every aspect of whisky production. When he returned to his studies in Glasgow, he took lodgings and grew friendly with a Jessie Cowan, and less than a year later, despite the reservation of their respective families, they were married at a registry office in Glasgow. Eventually the young couple found themselves in Japan, where Taketsuru, acting on the encouragement of his wife, set up the Yamazaki Distillery in partnership with a man named Shinjiro Torii. The rest, as they say, is history.

> *Whisky making is an act of cooperation between the blessings of nature and the wisdom of man.*
>
> Masataka Taketsuru

Japan is now considered an established whisky-making region and the country is known for producing quality expressions that can stand alongside the best in the world, with single malts from distilleries such as Hakushu, Suntory, Chichibu and Yamazaki setting the whisky world alight and cementing the country's status as a serious player.

Another leading light of New World whisky is Taiwan. The island boasts two world-class distilleries in Nantou and Kavalan. The latter in particular is hugely successful, with its Scots-inspired single malts being much sought after by whisky enthusiasts around the world. The whiskies produced here are distilled in Scottish pot stills and use the traditional methods that have been practised in Scotland for centuries, but with enough of their own twists to make their spirits deliciously unique. In 2015 this resulted in the Kavalan Solist making headlines across the globe when the expression was named the best whisky in the world at the World Whisky Awards. The country is also the world's third biggest consumer of single malt whiskies.

However, the biggest market for single malts in the world is India, and although its own whisky production is still in relative infancy, it is certainly thriving. At the time of writing, the country boasts more than a dozen single malt distilleries, with perhaps the best known being the Rampur and Amrut distilleries, both of whom have released well-regarded expressions over recent years. The Indian climate allows for much quicker maturation than say Scotland or Ireland – it is believed that one year of ageing in these countries is similar to three years in the much colder climates – and this has resulted in expressions from the country gaining notice on the world stage. Much of the whisky produced in the country is on the young side but with many casks ageing slowly there is little doubt that the country will before much longer, if it has not already done so, develop a whisky that can compete with anything that the more established whisky-making countries produce.

Australia is another huge player on the world whisky stage, with the country housing more than a hundred distilleries that produce the spirit. Once again the whisky produced here very much follows the Scottish process, but with its own distinctive style that results in a spirit that has a taste that is unique to the region. The Australian style was hugely popularised when in 2014 Sullivans Cove Oak single cask took the best single malt trophy at the World Whisky Awards. This was the first time a distillery from a country other than Scotland or Japan had taken the accolade.

From big to small though, and even tiny countries such as my own native Wales are making their presence felt. When Penderyn, situated at the foothills of the majestic Brecon Beacons, released its first single malt on St David's Day 2004, it became the first Welsh whisky distillery to release its own product in over a century. So successful has the distillery become that there are now several more Welsh distilleries, each producing a style of malt whisky that is uniquely Welsh.

Welsh whisky, although an infant on the world stage, holds a history that stretches back deep into the history of the drink itself. In 1705 Evan Williams left his native Pembrokeshire for the United States, where he founded a distillery in Kentucky and became a founding father of the bourbon industry. However, back in Wales there were entrepreneurs who believed they could produce a spirit that would challenge the dominance of the Scottish and Irish. In 1889 the Frongoch distillery opened its doors, but due to a strong temperance movement and considerable difficulty in reaching a large market, the distillery closed down in 1903 and Welsh whisky was effectively dead. That was, or course, until Penderyn released its first bottlings more than a century later and revived the ghost of a long-dead dream to produce a quality whisky that was very much a unique product of Wales.

Penderyn has gone from strength to strength in the decades following and now operates several distilleries across South Wales, but its first distillery will always be the spiritual home of the brand. It was from the tiny village of Penderyn that the brand took its name, and it is from nearby that the water is sourced that goes into the award-winning malts that the distillery has become known and celebrated for.

The first distillation was carried out in the year 2000 and this was ready for release four years later. Fittingly, the release was set for 1 March 2004, St David's Day, when the patron saint of Wales is celebrated across the country. The Welsh, for a small nation, are hugely proud of their cultural history and the team behind Penderyn were determined that their whisky would be uniquely Welsh. HRH King Charles, then the Prince of Wales, was on hand for the celebrations that saw the first commercial Welsh whisky in more than a century take tentative steps towards a crowded but ultimately receptive market and today the brand is exported to over fifty countries as well as serving a thriving domestic market.

When the first bottlings of Penderyn hit the stores, single malts were very much synonymous with Scotland. In some ways they still are, although this has become less so over recent years. And the team at the fledgling Welsh distillery had the foresight to realise that it was futile to even attempt pitting their small operation against the monolithic Scottish malt industry, and so they set about making a big noise over the unique aspects of the spirits they produced. Welsh malts were lighter than the typical Scotch malt but no less tasty for it. They were different and the first expressions the company released were all branded to make the most of the Welsh origins. In 2012 the company released what would be the first in their successful 'Icons of Wales' series with an expression called Red Flag.

The Red Flag bottlings commemorated the Merthyr uprising of 1831, an industrial conflict that saw for the first time in the United

The original distillery in Penderyn. *Author's collection*

Kingdom a red flag, actually a large rag soaked in the blood of a slaughtered cow, raised in the name of working-class revolution and social justice. That flag may have been the representation the company chose to illustrate the bottle label but it is the plight of one particular coal miner, Richard Lewis, AKA Dic Penderyn, who was executed on the flimsiest of evidence for his part in the industrial uprising, that is still remembered with a sense of injustice to the present day.

There is little doubt that the verdict passed down on Penderyn was a miscarriage of justice, but all the campaigns – there have been many – to secure a posthumous pardon have failed despite the efforts of many prominent politicians, and at the time of writing the campaign to have him recognised as an innocent man sent to the gallows remains very much alive. The town of Merthyr Tydfil, scene of the uprising, hosts the annual Merthyr Rising Festival, which celebrates working-class art, culture and performance, and once again the red flag flies in the memory of men like Penderyn, who dared to challenge the status quo of an unfair and unjust working environment. He remains a martyr for the cause of working-class solidarity.

The first expression in the Icons of Wales bottlings. *Author's collection*

Penderyn, along with another man, Lewis Lewis, had been arrested for the stabbing with a bayonet of Private Donald Black of the 93rd Highland Regiment, who were among the many army units brought in to subdue the working-class army who had mobilised themselves to protest against the lowering of wages and the high cost of bread. The stabbing did not prove fatal and Private Black was unable to identify either of the two men as his attackers. Nevertheless, the trial

The final resting place of a working-class hero. *Author's collection*

held at Cardiff Assizes found both men guilty and death sentences was duly handed down.

Following a statement from a special constable who said that Lewis Lewis had defended him from the rioters during the violent scuffle, Lewis had his sentence commuted to transportation. However, the authorities knew that they had to present a show of strength and send out the message that riots, no matter how just the cause, would not be tolerated. They were determined that at least one man should die and so on 13 August 1831, Penderyn was led to the gallows and lost his life in a public execution. The message from the authorities seemed to have little to do with actual justice but rather one of 'know your place'.

'*O Arglwydd, dyma gamwedd,*' Penderyn said as he stood upon the gallows, looking down at the gathered crowd. This translates as 'Oh Lord, here is iniquity.'

Penderyn is no longer alone and several more Welsh distilleries have come into existence and found their own place in the market.

Da Mhile makes use of an ancient farm to produce its excellent whisky. *Da Mhile Distillery*

Da Mhile (Gaelic for 2,000), situated in the picturesque Ceredigion valley, is particularly interesting and in 1992 was the first distillery of the modern age to produce an organic whisky. The distillery was the brainchild of John Savage-Onstwedder, who saw a gap in the market. The organic whisky was distilled at Springbank, Scotland, and the single malt was a one-off, bottled to celebrate the new millennium. So successful was it that Savage-Onstwedder worked with the Loch Lomond distillery to produce the world's first organic grain whisky. From there the businessman and environmentalist decided to open his own distillery. And so Da Mhile was born, which helped establish Wales as an official whisky region – European Union legislation states there must be at least two distilleries in a country for it to be recognised as a region, and of course Penderyn was the first.

Da Mhile Single Malt Whisky

> *Maturation in sherry casks have given the whisky a beautiful golden hue, and a nose that teases fruit, cinnamon and butterscotch.*
>
> *On the palate it is incredibly complex, and tastes like a much older whisky with salted butter notes leading the charge, with treacle and caramel coming through.*
>
> *It finishes well, lingers long as the buttery taste fades and reveals toasted walnuts to end.*

In 2009, England joined the other home countries when the first bottlings were released from the English Whisky Company, which had established itself in 2005. These bottlings were the first English whisky in more than a hundred years, and the success of the company has seen others take up the baton with their own take on an English whisky style. While looking towards the Scottish method for inspiration, these whiskies are very much their own thing, and the English Whisky Guild, a trade body set up to promote English whisky, now has more than twenty distilleries as members.

Cotswold Single Malt

Matured in both bourbon and sherry casks, this whisky is a deep golden colour, with a spicy cereal aroma.

Sweet citrus flavours dominate but there is some oaky spice and a nice creamy mouthfeel.

Porridge and some pepper on a truly lingering finish.

Chapter 11

How Whisky is Made

It is a basic truth that whisky is made from three things: grain, water and yeast, and although there are variations in grain types and quantities of each specific grain used depending on the style of whisky being produced, this truism remains whether a person is drinking a single malt, a rye or any of the other variations that these days grace whisky shelves. Whisky making may, in one sense, be an exact science using knowledge gained across the centuries, but it is also an art form, with many of its secrets shrouded in mystery and wonder.

In this chapter we will concentrate on the Scottish single malt style of whisky, but the production technique basically remains the same across the board. Barley is the main ingredient used in single malt and the first process on the way to becoming a glass of amber gold is for this to be malted.

> *Whisky is made from malted barley and the morning dew on angel's nipples.*
>
> Warren Ellis

Traditionally malting was back-breaking work that was originally carried out by hand at individual distilleries, though these days, save for a few instances where the old ways are retained largely to please the whisky tourists, this is carried out in huge industrial facilities that serve the needs of any number of distilleries. The old style of floor malting consisted of the barley being steeped in water, increasing the moisture content of the grain from around 12% to 40%. The grain was then laid out on the floors of distilleries to germinate, with the grains turned regularly by men with large wooden shovels. The shifts would last for many hours with the men continuously turning tonnes of barley in order to achieve a consistent germination. Over time this heavy work would bring on a form of repetitive strain injury known as monkey shoulder, and workers suffering from this infliction would find one of their arms hanging down very much like that of a chimpanzee. To add to this temporary infliction, there were also the widespread and far more serious respiratory problems reported, which were a direct result of the high levels of dust in the malting rooms.

By the 1970s the vast majority of floor maltings were abandoned in favour of mechanical malting, with the once common monkey shoulder being commemorated by a popular blended Scotch that carries the name.

The barley being assessed at the Cotswold Distillery. *Cotswold Distillery*

Monkey Shoulder blended malt whisky. *Author's collection*

The aim of the malting process is to trigger the germination of the barley seed by providing a warm and moist environment, tricking the seeds into spring growth before stopping further development by drying the seed with some sort of heat source. This partial germination ensures that the grain is modified so that the starches and proteins are in a form suitable to provide the extract that the distiller needs. In some cases, the heat used to dry the now malted barley will include peat and it is this decayed vegetation that has broken down over many thousands of years that provides the finished whisky with that much-beloved smoky character. These days the peated style of whisky is most associated with coastal malt whiskies but in years gone by peat was used in the kilns of the majority of distilleries. Indeed, the reputation of Scotch whisky was originally characterised by its smoky quality.

To go off on a tangent for a moment, it is worth pointing out that the extraction of peat harms vital eco systems, and it is hoped that its use in whisky will be totally phased out by 2030. Already, a high-profile piece of legislation has seen peat banned for use by domestic gardeners and a similar move for its use in whisky is inevitable. Already, distillers are experimenting with alternatives in the kilning process, the drying of the barley, such as juniper, birch bark and even sheep's dung. Going forward, this should provide some interesting tasting notes.

The next step on the road towards the finished product is the milling and mashing of the malted barley. This process sees the dried barley milled into a coarse mixture called grist and this is made up ideally of 70% grit, 20% husk and 10% flour, although there are often small variations but such mixtures are considered to have the ideal proportions by the distilling industry. This process is carried out with, alongside modern equipment, a contraption called a shoogle box, which separates the three elements in the grist when shaken, or shoogled to use the correct terminology. There are many safeguards during this stage in production, with the malted barley first being passed through a device called a dresser to remove any stones from the mixture, while a powerful magnet ensures that any alien metal objects are also removed from the mix. Once that is all done it is time to move onto the next step, which is the mashing of the grist.

The grist is then transferred into a mash tun, basically a vessel that mixes the grist with temperature-controlled water. What started out as barley is now a porridge-like mixture and this is turned within the tun, while the starches in the malt now convert into fermentable sugars. The substance is then treated to two more bursts of increasingly hot water, while a final third of the water is kept back to become the starting point of the next batch of grist. The residue of this process is rich in protein and is a traditional animal feed, which is increasingly processed into pellets and sold to animal food suppliers.

The grist has now become wort, a sugary liquid. Once cooled, yeast is added and it is placed in the tun room; it is now that fermentation takes place. Large wooden containers, or tuns as they are known, are used to ferment the wort to produce what is known as the wash. This is basically a beer of around 8% ABV or alcohol by volume. These days it is not unusual for the tuns, or washbacks as they are more commonly called, to be made from stainless steel rather than wood. Traditionalists, though, hold to the view that the wooden tuns, washbacks, play a part in developing flavour in the final spirit and because of this some distilleries still use them. The veracity of the last point – wooden washbacks being better than stainless steel ones – is hotly debated by whisky experts, but there is no doubting that the length of fermentation has an effect on the taste of the final whisky. A fermentation of only a couple of days is said to impart a nutty flavour, while a longer fermentation brings out more complex, fruity notes. The scientific explanation for this is that the presence of lactobacilli, a bacterium, helps to develop flavour and that the longer the wash interacts with the bacterium the more pronounced the flavours become.

The final stage is the distillation of the wort, and although this process is quite simple, it is overburdened with distillers' jargon that only serves the confuse the layman. However, stills at their most basic are large copper vessels within which the wort is boiled in order to gather the spirit from the condensed vapours. Distillation is carried out several times and most stills usually operate in pairs, so a distillery may have two or four, six or even more but there are rare

Fermentation occurring at the Pearse Lyons Distillery. *Grace O'Neill, Pearse Lyons*

distilleries that run an odd number of stills. Laphroaig, for instance, has seven but an odd number of stills usually suggests an unorthodox distillation regime that may or may not deliver a specific character to the finished whisky.

There are basically two types of stills used in whisky production, and although there may be variations on the theme, the most common stills are the pot still, which on its first run produces a whisky of around 20% ABV, and so the process of boiling and condensing is repeated several times until a whisky of somewhere between 60 and 80% ABV is produced. The other type of still is the column still, also known as a Coffey still, and these are more effective than pot stills because they operate a continuous process that removes the need to run several distillations. The wash, or wort, is poured into the top of the still and passes through perforated plates, while hot steam rises from the bottom of the still. This process can produce a whisky of around 95% ABV but most distillers aim for a lower proof.

So that's it, the finished whisky. Well, not quite for the most important step is still to come and it is undoubtably this process that gives the whisky most of its character. Whisky straight from the still is clear, like vodka, and it is from its time in the barrel, or cask, that most of its flavour and colour comes. This maturation is an essential element, not to mention a legal requirement, of whisky production in many territories.

Scotch whisky, for instance, must be aged for a minimum of three years and a day in the barrel before it even qualifies to be called Scotch, while Irish whiskey requires a bare minimum of three years and straight bourbon must have two years in the wood. For the latter whiskey even the type of barrel is specified, with bourbon having to be aged in brand new American white oak of which the interior has been charred with a flame prior to the spirit entering. Recently laws were passed in Japan that stated their whisky must, like Scotch and Irish, have at least three years in the barrel.

The time spent within the cask is vitally important, and the seemingly magical interaction between the wood and spirit is a truly mysterious alchemy – the spirit goes in harsh and crystal clear, and due to time spent interacting with the staves of the barrel emerges as a smooth amber gold liquid. Whisky experts universally agree that at least 60%, some claim it to be even higher, of the drink's final flavour and the totality of its appearance, unless colourings are added later, can be attributed to the time it spends ageing within the wood.

Let's look at the barrels, or casks as the Scottish like to call them. An expertly constructed whisky barrel will contain no nails or glues to hold it together, but will consist of a series of staves, pieces of wood that have been sawn from the oak and then shaped to form the sides of the barrel. These staves will have a bevelled edge and are held together by metal hoops. These barrels will have been made in a cooperage and those who actually construct the barrels are called coopers. The word cooper is thought to have originated from the Latin word *cupa*, which means vat.

The oak used to have constructed the barrels will have been carefully selected for its strength and ability to remain liquid tight,

Whisky sits in the casks for several years. *Cotswold Distillery*

though porous enough to allow air in, when coopered together. The sizes of these barrels will vary greatly but in Scotland the most commonly used kinds are American standard barrels, which hold 190 litres; butts, which hold 500 litres; puncheons, which hold 550 litres; and hogsheads, which hold 250 litres. Smaller barrels are sometimes used to speed up the maturation process, such as quarter casks. It is believed that a whisky that has been aged three years in a quarter cask will have the characteristics of a whisky aged for a much longer period in a more conventionally sized barrel. This is believed to be because of the greater interaction between the whisky and the wood in a smaller barrel.

Over time the oak will extract unwanted elements from the new whisky while giving it desired characteristics such as vanillins, lactones and tannins, which all add flavour and colour. There have been hundreds of flavour compounds identified in whisky and, as we have explained previously, the bulk of these come from the time the whisky has spent in the wood. The original variety of grain or the water used in production are only a small contributor to the liquid in the glass that tantalises the taste buds of consumers everywhere. Perhaps an exception and a factor that does carry over to the taste of the finished whisky is the use of peat in the malting process, which releases phenols into the barley, and peat from different regions will impart quite unique levels of smoky qualities to the eventual taste of the whisky. However, as noted earlier in this chapter, the use of peat in whisky production is expected to come to an end by the end of the decade.

There are those distillers, however, or rather marketing men who have tried to introduce the concept of *terroir*, a French term much used by the wine industry, into whisky making. The concept is undoubtably important when taken in relation to wine making with the grape variety, the climactic area and even the weather conditions during the year of harvest being a vital factor in the eventual taste of the wine. However, whisky is distilled not only once, but twice or even thrice to a purity that would ensure the effects, if any, of the earth, weather conditions and climate of where the original grain was grown will be at best minimal. The same thing applies largely to the source of the water used to start the fermentation; even if distilleries do make a big deal of their water usually being sourced from local springs, this is unlikely to have any effect on the finished drink. It is laughable in fact when it is taken into account that often two of more distilleries may use the same water source and yet each will produce a unique whisky of their own, while they continue to point to the uniqueness of their water properties.

If *terroir* is indeed applicable to whisky production, then the biggest factor is the barrels used to mature the drink because in terms of whisky and its flavour, even its appearance, wood is very much king.

While the barrel cannot make a bad whisky good, it can make a good whisky great, but badly made barrels can produce a rather bland whisky. It is the wood that makes the whisky, is a mantra that was often heard, and indeed still is, within the industry. When this is thought about in detail it seems only logical since it takes but a matter of a week at the very most to transform whisky from unmalted barley to the liquid filling the barrel. There then follows a much longer period spent maturing – remember three years is the minimum for Scotch – within the confines of the barrel. There was a period when Scotch could spend its years ageing within any kind of wooden barrel, only that barrel had to be wooden was specified in law, but now only oak can be used. It is surprising that this only became law in 1988, though in practice oak barrels had long been favoured by the Scottish whisky industry since oak is so strong and allows the spirit within to breathe as it matures, thus creating the interaction between spirit and oxygen.

The four major components of oak are cellulose, lignin, hemicellulose and tannin, and these elements can add sweetness and colour to the finished whisky as well as subtracting a lot of impurities. And because Scotch is matured in previously used barrels, just because it is considered better to do so, other factors come into play. Residual flavours left over from whatever the barrel previously contained will pass over to the finished whisky, which is why producers often make a point of listing the barrels used for the maturation of their whisky in their advertising.

By far the most commonly used barrels in Scotch are those that once contained bourbon, but used sherry barrels are also extremely popular as well as those that have contained other wines. The first time these barrels are filled with whisky they are called first fill and thereafter they are termed refills. After several refills, when the staves of the barrels are considered to have been spent and have nothing else to contribute to a whisky, they are broken up to be used as sold fuel or sold off to garden centres to be adapted into ornamental planters.

Ahh, the wonders of wood!

Chapter 12

A Double Whisky Mixer

James Bond: Whisky is Forever

With the whisky industry worldwide in such rude health, it is easy to forget that during the 1960s and 1970s sales were in decline right across the board. It has been theorised that part of the reason for the downturn was that the drink had failed to move with the times and had suddenly become staid and unhip. It was being seen as a drink for the old folks, while the young wanted more variety, more colour, more sparkle.

Much of the blame for this perceived lack of cool was placed at the door of James Bond, the fictional icon who for much of this time seemed to personify coolness itself. His tipple of choice, we were often told, was a vodka martini, shaken not stirred. Yet in the original novels Bond often consumes whisky or even whiskey. In *Live and Let Die*, the debonair spy orders an Old Fashioned cocktail made with Old Grand-Dad bourbon and he repeats this drink in *Thunderball*. Often he simply orders a generic Scotch but in the novel *You Only Live Twice* a bottle of Suntory features.

The literary James Bond did indeed enjoy a tipple or two of the amber gold, but when it comes to the movie version whisky has largely been conspicuous by its absence, and fans everywhere started ordering vodka martinis, not forgetting to specify shaken not stirred. It is odd that Sean Connery, a Scottish actor playing a part-Scottish character, is seldom seen with a glass of the national drink. It is only in *Goldfinger* that the amber gold makes an appearance and even then it was bourbon rather than Scotch. It wasn't until the Pierce Brosnan era that Scotch actually made an appearance in the film series in the shape of a bottle of Talisker in *The World is Not Enough*. From there it was not until Daniel Craig's third Bond, *Skyfall*, that whisky made a notable comeback in the world of 007. The movie featured several scenes that featured a 50-year-old Macallan, with the most memorable being the incredibly tense one in which Bond is forced to use the whisky for target practice.

So, did the public fascination with James Bond cause a dip in the health of the Scotch industry? It may have been a contributing factor, but a small one at that, in what was already a downwardly looking market. However, doom would soon turn to boom and today the whisky industry, Scotch included, has never faced a brighter future. So, what if the celluloid James Bond was not a whisky drinker, we known that the literary version, arguably the real thing, was very much a man of the amber gold. Below is a list, as near enough as I can make it, of every kind of whisky Fleming's Bond drinks in the series of original books. If I've missed any out then I apologise but it's an arduous task flicking through the series of novels looking for mention of whisky. After all, unlike Bond, you only live once.

Bond drinks Haig and Haig Pinch in both *Live and Let Die* and *Moonraker*, and again in *Live and Let Die* Bond also downs a glass of Old Grand-Dad bourbon. So fond of this particular bourbon was 007 that he even specified it when ordering an Old Fashioned in *Diamonds are Forever*. Bond also takes Black and Black Scotch in *Moonraker* but by *Dr No* he had switched his allegiance to Canadian Club blended whisky. Once back *On Her Majesty's Secret Service* Bond is again drinking bourbon, this time a bottle of I.W. Harpers, and in the same novel he enjoys a tipple of Frank Sinatra's favourite pour, Old Jack Daniel's No. 7. In the penultimate Fleming novel, *You Only Live Twice*, Bond partakes in Suntory Japanese whisky and is pleasantly surprised given the snobbery he earlier displayed about whisky made in Japan. However, by the next novel, the posthumously published *The Man with the Golden Gun*, Bond is back on the bourbon trail with Walker's Fine Old Bourbon Whiskey.

The Great American Fraud

In 1898, America once again found itself at war when President William McKinley declared hostilities against Spain. The spark that lit the fuse of the conflict occurred when the Spanish blew up the American battleship USS *Maine*, which had been sent to help Cuban gorillas in their conflict with Spain.

'Remember the Maine', screamed the headlines of American newspapers.

Congress had a budgetary problem, though and with coffers being so tight it was difficult to see how the war could be bankrolled. Once again, the politicians looked to tax whiskey but this time it wouldn't be just alcohol that would be subject to taxation but patent medicines too. The taxes placed on the medicines were obviously set at a significantly lower rate than that on basic alcohol.

New York businessman and general raconteur Walter Duffy, rather than cursing the increased taxation, actually saw this as a great opportunity. After all, he had long marketed his Duffy's Pure Malt Whiskey as a cure-all tonic, a magical elixir for all manner of ills, and each bottle actually came with a dosing spoon. Several ailments – consumption, bowel problems and dyspepsia – were all specified on the label and many more were claimed in the advertising. None of these claims were credible but that didn't matter to a man like Duffy. He knew that the possibilities would be endless if he could only secure a federal tax stamp for his whiskey.

Duffy proceeded to furiously lobby Congress for his beverage to be given the status of a patent medicine and he got his way, largely because the word malt, a healthy substance if ever there was one, on the label carried such sway. For many years, Duffy had been making outlandish claims about the health benefits of his whiskey and now, at long last, this seemed to be justified with it being categorised as a patented medicine, and carrying the federal seal of approval. The bottom line was that Duffy could now save thousands of dollars due to the reductions in taxation that the tax stamp brought him.

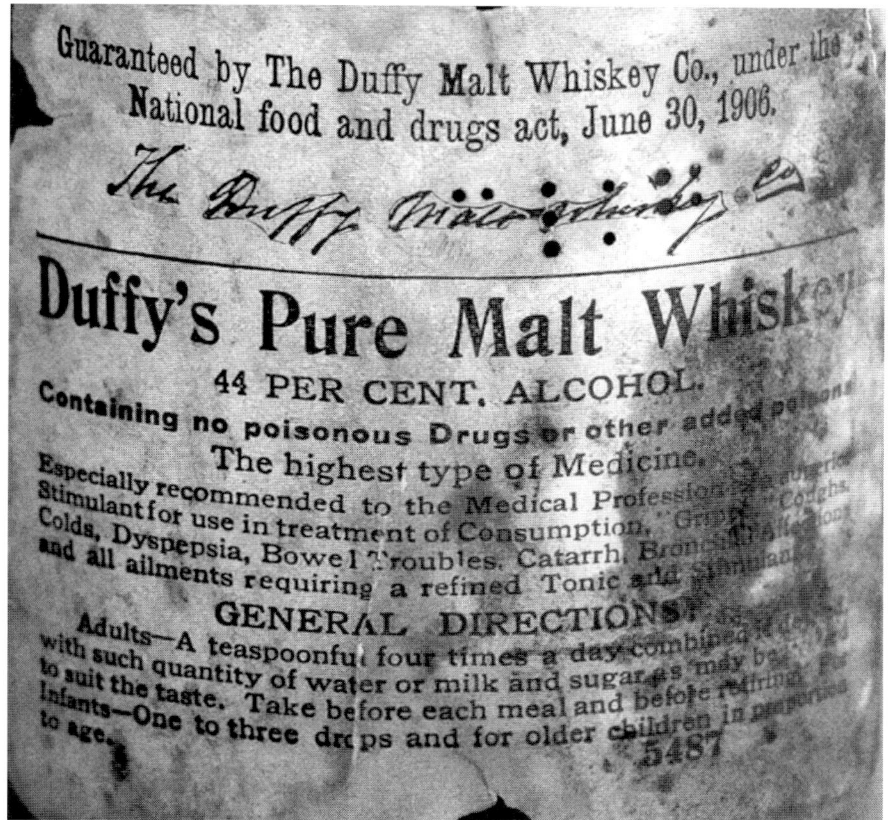

Duffy's Pure Malt. *Author's collection*

Duffy now stepped up his advertising, something he had never been afraid of, and began to extol the rejuvenating powers of his pure malt whiskey.

'Why grow old when you can be strong and vigorous at 100 years old,' asked one newspaper advertisement.

Duffy even backed up these claims by publishing testimonials, one seemingly from a 120-year-old man who claimed he had spent the past few decades drinking Duffy's as his only medicine and felt as chipper as he had in his forties. Another much younger man, a Mr Townsend, claimed that at 104 he felt as fit as any 25-year-old, and the testimonials all likely authored by Duffy himself continued. The secret presumably known to these old age imbibers was, or so the adverts claimed, because Duffy's whiskey contained no fusel oil, a chemical that is mostly removed from whiskey during distillation and was a pure stimulant.

'Absolutely pure, makes the weak strong', went one particular advertisement, while another advised protentional health-conscious customers to, 'Drink Duffy's Pure Malt when you are not feeling well'.

'I really don't feel a day over sixty, thanks to Duffy's Pure Malt. That is the secret to my longevity,' a 106-year-old Nancy Teague was quoted in an advertisement that ran in *The Minneapolis Journal*. The article even carried a picture of the care-free woman.

It will seem incredible to the reader, but Duffy was allowed to carry on with this sort of advertising for many years; with the federal seal of approval adding legitimacy to his outlandish claims. Duffy knew this and was quick to point out that his whiskey was the only one approved by the government for use as a medicine.

'The absolute purity of Duffy's can be attested to by the fact that thousands of doctors and hospitals have prescribed it to cure all manner of ills,' he boasted shamelessly whenever the veracity of his claims were brought into question.

Duffy continued to prosper and even a 1906 speech to Congress by James Mann of Illinois in which he listed the many cases of poisonous chemicals being added to whiskey did little to slow down sales. However, this was to change when a series of articles were published in American newspapers that took direct swipes at Duffy and his so-called miracle tonic. A journalist even managed to track down the 106-year-old Nancy Teague and the old woman, practically blind, living an old people's home, spoke to the reporter. She was aware of the advertisement that carried her likeness but claimed that she had been tricked into both being photographed and signing the testimonial by Duffy, and what's more she didn't touch alcohol of any kind, and never had. She would eventually die one year later at the age of 107 and is buried in St Mary's Cemetery in Lafayette.

Still Duffy seemed untouchable and he won lawsuit after lawsuit but eventually the New York Commissioner of Excise, Patrick

An example of outrageously false advertising. *Author's collection*

Mulligan, took Duffy to court over his claims. His whiskey was nothing more than a sweetened beverage with no active medical ingredients, the court was told but Duffy fought back and brought no fewer than eleven physicians into court and each of them swore that the whiskey did indeed contain drugs. The case snowballed and ended up in the Supreme Court, where the state eventually won the case against Duffy, but that was just a minor irritation to the businessman. That was just one state out of fifty.

Duffy would continue with his wild claims of the health-giving qualities of his product until his own death at the age of just 70 in 1911. Maybe he should have drunk a little more of his own whiskey. His son, also called Walter Duffy, took over the business, but in 1915 the company was forced to remove all medical claims from the bottle labels with only the word tonic being allowed to remain. A combination of increased pressure from the Food and Drug Administration and American prohibition would soon see the Duffy brand consigned to history.

Shackleton and the Return of an Iconic Whisky

Born in Kilkea, County Kildare, Ireland, in 1874 but raised in London, Ernest Shackleton was a famed polar explorer who is even today still talked about in awe. It was on one of his many expeditions, a failed attempt to reach the South Pole, that he left behind something that would turn out to be a gift to future whisky enthusiasts around the world.

When Shackleton and his team were forced to abandon their Antarctic camp and return home in 1909, they left behind three cases of whisky that would be recovered from the ice more than a century later and thanks to this find, and the expertise of modern blenders, there is now a blended malt Scotch whisky on sale called Shackleton's. A selling point of this whisky is that is as close as possible a recreation of the whisky Shackleton and his team carried with them on the British Antarctic expedition all those years ago.

Shackleton and his team had reached a point where they were only 97 geographical miles short of their target of reaching the South Pole when they were forced to turn back due to dwindling supplies and horrendous weather conditions. It was clearly unsafe to continue, the men had been pushed to the limits of human endurance and Shackleton, in fear for his team's lives, made the difficult decision to abort the attempt. The Nimrod expedition may have failed in its objective but many records had been set, including reaching the then furthest southern latitude, and Shackleton was knighted by King Edward VII upon his return. In fact, all the men in the expedition were later awarded medals by the Royal Geographical Society.

'The only thing that stopped us was the lack of fifty pounds of food,' Shackleton told reporters on his return home. 'We made many interesting geological and scientific discoveries and had many narrow escapes throughout our journey.'

In 2010 a team of conservationists from New Zealand's Antarctic Trust found three cases of Mackinlay's whisky that had been left behind when the Nimrod expedition had to abandon their camp in the Antarctic. The whisky had spent more than a hundred years in the ice, but due to its high alcohol content had not actually frozen and within the bottles it sloshed around like a regular whisky. The conservationists reached out to Whyte and Mackay, the present-day owners of the Mackinlay brand, and in 2011 a private jet flew several bottles from Christchurch, New Zealand, to Scotland for careful analysis. Would the whisky be safe to drink? Although whisky only ages within the barrel, this was truly aged whisky. It had been bottled at 15 years old, so in actuality it was still 15 years old, and yet it had spent more than a hundred years in a frozen wasteland. It was 15-year-old whisky in 100-year-old bottles.

To go off on a tangent for a moment, and for the benefit of any readers who don't fully understand the concept of whisky ageing, it is quite simple. Whisky only ever matures within the barrel, that is changes, alters in both taste and appearance, becomes something else, due to the influence of the wooden barrel within which it rests. Once the whisky is bottled it ceases to change, no further maturation is possible and so in effect a whisky that is bottled after ten years in the barrel remains 10 years old throughout infinity. Readers who have been paying attention will know this from the earlier section in this book that dealt with the importance of ageing new spirits within a wooden barrel or cask. A particular whisky that was bottled after ten years within the barrel may not actually be opened for a billion, trillion, squillion years, and drunk by some future evolution of the human species, *homo sapiens divinus*, perhaps. However, it will remain forever 10 years old. That's all easily understandable – come on, it's far simpler to grasp than the whole whisky, whiskey, *uisge beatha* thing! But let's not go there again.

> *I wanted to capture the essence of Sir Ernest Shackleton when recreating this whisky.*
>
> Richard Paterson, head blender at Whyte and Mackay

Master blender Richard Paterson, known affectionately as The Nose due to his well-developed olfactory skills, was able to withdraw a little whisky from one of the aged bottles by carefully piercing the cork with a syringe. The bottles couldn't be opened, and had to be treated with the greatest of care. The labels, the cork, the bottles themselves all had to be preserved since the Swedish Antarctic Act of 2006 specifically makes it an offence to remove organic or inorganic materials from the continent of Antarctica, and the bottles would eventually have to be returned there, where they would be placed back into their icy tomb beneath the hut that had served as a base camp for the Shackleton expedition. It is there that they, alongside the other bottles that had been discovered, remain today.

Paterson and the team from Whyte and Mackay were aware of the level of trust that had been placed in them by allowing them to bring the ancient whisky to Scotland so that it could be analysed and recreated with minimal disturbance to the actual bottles of whisky, which had now taken on the status of priceless historical artefacts. It had taken a lot of careful negotiations involving not only the governments of New Zealand and Sweden, but the Antarctic Trust and Whyte and Mackay before the whisky, placed in secure cases and handcuffed to Richard Paterson, was allowed to be flown by private jet to the Whyte and Mackay headquarters. An onlooker, witnessing the level of security as the flamboyant Scottish master distiller and blender made his way to the private jet, would have no doubt thought the case handcuffed to his wrist contained jewels, money or even sensitive government documents rather than several bottles of whisky.

> *Even holding the bottles was especially poignant and emotional.*
>
> Richard Paterson

Analysis revealed that the whisky was 47.3% ABV and was lightly peated, with the peat used in the original malting having originated from the Orkney Islands. That it was so lightly peated was a surprise since nineteenth-century whiskies had usually been heavy, carrying much more of a peated, smoky flavour, but this was surprisingly light in style. The whisky had been matured in American oak casks that had previously been used to hold sherry or a thick wine, and it had been distilled at Glen Mhor, a distillery that had closed down in 1983 and been demolished several years later. Armed with this information, Paterson was able to recreate this whisky and two different expressions were created, both using the exact technique and where possible the same ingredients to those used all those years ago. The aim was to create a whisky that would have been recognisable to Shackleton, a taste that the explorer and his men would have been familiar with.

The first expression to be released was called Journey, while the second went under the title of Discovery. A percentage of the profits from sales of both whiskies went to the Antarctic Trust, but so successful, not to mention expensive, were both offerings that Paterson decided to create an affordable mainstream blended malt whisky simply called Shackleton, which while not an exact replica is in the same style as those ancient bottles that had been discovered in the ice. This meant that whisky consumers were able to taste a style of whisky that stretched back to the nineteenth century. It remains a popular blend and a strong seller to this day.

Rather than call this a recreation, Paterson insists that this a personality blend, saying he wanted to capture the essence of

Shackleton, a taste of history. *Author's collection*

Sir Ernest Shackleton in the whisky. He saw Shackleton as a man of great courage and fortitude, so he used a lot of the stronger Highland malts as the backbone of the blend. However, Shackleton was also a lover of women and so lighter Speyside malts were added, making the whisky far more accessible to casual drinkers. It was bottled at the standard 40% ABV, rather than the rather potent 47.3% of the original Mackinlay upon which it was based. This makes it comparable to the everyday blends and single malts that stock supermarket shelves up and down the country.

Shackleton's Malt Whisky

> *The malts used in this blend are aged in a mixture of ex-bourbon and sherry barrels, which gives it a rich golden colour. It's sweet on the nose with hints of creamy toffee and a faint wisp of smoke.*
>
> *Honey sweetness on the palette, with vanilla and woody notes with some aniseed.*
>
> *The finish is long with traces of peat and citrus.*

A Scottish Funeral, Far Merrier than any English Wedding

The phrase 'to be late for one's own funeral' is believed by many to have originated from the Scottish tradition of toasting the recently deceased with several drams of good-quality whisky. The phrase has popped up in print several times throughout the years, with perhaps the earliest example of its use coming from an article by George Finch Mason, a sporting journalist who in an article published in *Bell's Life in London and Sporting Chronicle* in 1881, claimed that a young gentleman named Charles Wildoats was so lackadaisical in his punctuality that he would no doubt be late for his own funeral.

However, as we have discovered, the history of whisky is full of legends, myths, half-truths and even downright lies. We have already seen that the curious reader who delves into whisky history will learn of ghosts, ghouls, ferocious creatures, in fact all manner of mythic wonders. Even the devil itself, that demonic monarch of nightmares, crops up in several well-known whisky legends. With so much fancy attached to whisky history then it will come as no surprise to learn that many whisky historians have laid the claim that the aforementioned phrase originates from the dim distant days of our beloved drink. What is surprising though is that there is a good deal of compelling, admittedly anecdotal, evidence for these claims.

The area of Angus was during the eighth century the territory of the Picts; indeed, the area takes its name from King Angus Macfergus, who led the Picts from AD 729 until his death in 761. Historically a province, it has over time been a sheriffdom and a county. Perhaps its greatest claim to fame is the world-famous beef produced in the area, though in terms of whisky, as well as being home to several major distilleries, it is also linked to the legend of a popular schoolteacher Jessie Colquhoun and it is she, it is said, who was indirectly responsible for the phrase 'to be late for one's own funeral'.

> *Love makes the world go round? Not at all. Whisky makes it go round twice as fast.*
>
> Compton Mackenzie

The Scottish tradition is for the memory of the recently deceased to be toasted with the finest whiskies, and such was the case with the service for Colquhoun. Such was the celebration that when the time came to carry the coffin to the kirkyard, which was some 4 miles away, her brother, Captain James Colquhoun, and the rest of the mourners were more than a little unsteady on their feet. Nevertheless, they lifted the coffin and set out on their journey. As was tradition, they stopped at every inn they came upon, laying the coffin down upon a leckerstane, a large stone that had been laid at the roadside for the use of funeral processions. At each inn they would drink several more toasts, and tell stories of the recently deceased.

When the procession eventually reached the kirkyard, they were devilishly late and Captain Colquhoun apologised to Ault Tam, the gravedigger on duty, for being so late. The gravedigger was said to have been annoyed to have been kept waiting, but was puzzled when he could see no coffin. Where was it, he asked and the captain and the other mourners all looked to each other, having realised that they had left the coffin upon the leckerstane outside the last inn they had visited. The men, perhaps a little sobered with embarrassment, turned back and retrieved the coffin and the schoolteacher was properly buried; and the language was enriched with the phrase 'to be late for one's own funeral'.

This story is well known in Scotland and beyond, and although there is no concrete evidence of its validity, it is an entertaining tale to tell while sharing a dram with like-minded acquaintances and friends. That's good enough for me.

History, Legend and Fable

Scotland alone has more than 300 working distilleries, and it is no surprise that over the centuries many stories and legends have become attached to whisky making, and almost every major brand has its own story to tell, which is pounced upon by the marketing men. With the water of life now being produced across the world, and enjoying rude health, having something that is unique to attach to a brand is highly desirable. And if there is one thing that marketing men have learned, it is not to let the truth get in the way of a good story. The consumers seem to like this too, and it is not uncommon for deluxe brands to carry highly fanciful folk tales, legends, truths and a smattering of bull-crap upon their packaging.

Several expressions of the Arran malt whisky carry an image of a majestic white stag upon its labelling, and the story of the beast is well known to the inhabitants of the area, with many sightings being reported across the years. The white stag is thought to be the leader of the deer population, a big and stable populace, but most of these are red deer and it is believed that anyone lucky enough to spot the white stag will have good fortune bestowed upon them. It is said that on the very morning that the Arran distillery opened, both the manager and head distiller spotted the fantastic creature looking down at them from the mists of the surrounding mountains. The distillery has thrived ever since. This story at least has more than a ring of truth as

Arran is home to an incredible array of wildlife from deer, to otters, seals, red squirrels and the fantastic golden eagle. Indeed, the actual building of the distillery itself had to be stopped temporarily in the 1990s because a pair of golden eagles were breeding in the nearby hills. This halt of work seemed to have worked and today the eagles can be seen soaring in the skies above the distillery. Now, that's a wonderful image to bring to mind as one sits back and appreciates the smoothness housed within a dram of the Arran malt.

Another expression from Arran is its Machrie Moor peated whisky, and the name of this whisky relates to the nearby moor, which is home to a group of neolithic standing stones known as the Stones of Arran.

The bottle label reads: 'The Peated Arran. Lightly peated at 20 ppm, this mythical malt has proved a popular addition to the Arran range. On the west coast of the Isle of Arran lies a windswept and mystical peat bog called Machrie Moor. Bronze Age stone circles and standing stones are strewn across its barren, undulating terrain. One of the stone circles is known as Fingal's Cauldron Seat, where sits a stone with a carved hole. The legendary warrior giant Fingal is said to have tethered his favourite dog Bran to this stone. This peated expression of the Arran Single Malt perfectly captures the rugged beauty and lore of the landscape. Unleash the legend that is Machrie Moor.'

The stones are believed to have been erected around 2000 BC, and there is archaeological evidence that suggests that they replaced a much earlier timber structure. It is believed these stones would have been used in religious ceremonies, the ancient pagans having worshipped nature itself, but later the stones circles were used for the burial of prominent members of the local populace, with both cremations and inhumations being carried out within the circles that the stones form.

Dalmore whisky is another brand that uses the image of a stag upon its label, this time a twelve-pointed royal stag. The story behind this is quite fascinating and dates back to 1263 when Colin of Kintail, the chief of clan Mackenzie, saved King Alexander III from being trampled by a crazed stag during a hunting trip. In gratitude for this act, the King granted the Mackenzie clan the right to use the emblem of the twelve-pointed stag on their coat of arms. And then when the decedents of the clan became the owners of the Dalmore distillery in 1867 it was an obvious step to use the twelve-pointed stag as the emblem of their whisky. Dalmore, of course, is quite a distinctive whisky due to its use of sherry casks from González Byass in Spain for the maturation of their whisky. A deal made more than a century ago between Dalmore and González means that these casks that had previously housed Matusalem Oloroso sherry are exclusive to Dalmore and give their spirits a delicious chocolate and orange character.

The Dalmore 12-Year-Old

Matured in white oak bourbon casks, and finished in Oloroso sherry casks.

Nose gives a rich roasted coffee, with a slight citrus tang.

A creamy mouthfeel with hints of dark chocolate, nutmeg and tangy marmalade that leads to a citrus heavy finish.

From regal stags to vengeful fairies, it really is a wonderful world of whisky.

Captain William Grant, not to be confused with he of Grant's whisky, may have become one of the most famous distillers in the history of Scotch had he not crossed the little people and brought ruination upon himself.

In 1809, having secured the rank of captain, he retired from the army and set up a farm in Auchorachan, Glenlivet, Scotland. The Glenlivet area was known as a locality where a lot of illicit whisky distillation had occurred during the eighteenth century, but when the process was legalised in 1823, the new distilleries that sprang up were hounded by the illicit producers and smugglers, who were angry that these new legalised concerns were taking away their trade.

There was much conflict and in 1826 James MacPherson, who owned the Tomalienen distillery, was ready to throw in the towel.

The royal stag of the Dalmore. *Richie Williams*

He had received both threats and attacks from many of the illicit distillers in the area, and had simply had enough. Captain Grant made MacPherson an offer for his business, and found himself the new owner of the distillery, which was soon going by the name of the Glenlivet Distillery and gaining worldwide fame, making good on Grant's boasts that he would soon be the most famous of all Scottish distillers.

The captain soon set to work constructing a new byre in the distillery grounds to house his cattle. However, he was short of a piece of masonry to form a lintel over the doorway of the new building and so he decided to take a large standing stone from the brow of a nearby hill. He was warned against doing to by distillery workers because the stone had stood there from time immemorial and was believed to have been placed there by fairies. There were local legends of evils having fallen on people who dared to offend the people of the fairy kingdom.

Grant though was a practical man and he simply ignored the warnings of the workers, telling them it was all superstition and hogwash.

The stone was quickly removed and installed in the byre and soon after the captain's cattle began to fall prey to a mysterious sickness. Cattlemen and veterinarians could not identify the illness and the animals continued to sicken and die. The local people advised the captain that the fairies had cursed his byre and that the cattle would continue to die until the stone was placed back in its rightful place, and finally at end of his tether the captain did just that. No sooner had the stone been removed from the byre and placed back in its rightful place than the mysterious sickness went and the cattle recovered and even thrived.

However, it seemed the fairy's vengeance was not yet done and Grant suffered bad luck for the rest of his life. One afternoon the captain's horse bolted, some say scared by several fairies that had run arounds its legs, and the captain was thrown. The local newspaper reported that he dislocated the joint at the top of his thigh and also fractured a bone there. The break would not heal correctly and forever after he walked with a severe limp and earned himself the nickname the Crippled Captain. When his wife died, the captain found that he could no longer manage his estate and had to sell up, distillery and all. He would himself die in 1877, penniless, crippled, a broken man, but the name of Grant would live on.

The Crippled Captain never did become the most famous distiller in Glenlivet, but he did leave his mark by way of his distillery manager, also called Grant but of no relation to the unfortunate captain. The distillery manager's son, James Grant, would learn the art of whisky making from his father and in 1895 he purchased the Highland Park Distillery on Orkney and his company, James Grant & Co., went onto become one of the legendary names in Scotch.

From little folk to decidedly big folk and the legend of how the Scottish drink Atholl Brose, a whisky liqueur made with oatmeal, honey and whisky, came to be. The ancient Pictish land of Atholl, now a part of modern-day Perthshire, was said to be terrorised by a giant who would steal cattle. This was bad enough but the giant would also rampage through grain stores, carrying away the grain in a huge sack and leaving the people of the area to struggle throughout winter with insufficient food. Many brave but foolhardy men set out to slay the giant, but none of them returned and their limbs would be found scattered across the hillside. The legend tells us that the giant ate a great portion of his victims.

That was until one day a brave and more importantly clever young man by the name of Dougal set out to find the giant. Dougal knew that his chances of taking him on were worse than slim and so instead he searched and searched for the giant's lair. One day he came across a large cave in the hills and luckily the giant was not home at the time. Dougal looked around and discovered several large sacks of grain, many jars of honey and two large whisky casks. Further examination of the cave revealed the giant's drinking cup, a huge hollowed-out boulder that rested beside a stone well. What followed was perhaps the first example of drink spiking and Dougal poured two sacks of oats into the giant's drinking vessel, then he added an amount of honey for sweetness and emptied both casks of whisky into the cup. Using a large stick, Dougal stirred the potent mixture and then found somewhere to conceal himself and await the return of the giant.

Soon the giant did indeed return, and Dougal watched from his concealment while the giant drank his fill and then lay back and fell into a deep whisky-infused slumber. After some time, the young man emerged from his concealment and using his sword he slit the throat of the still sleeping giant.

The young man returned home a hero, and with him came the recipe for the drink that would eventually become known as Atholl Brose; brose of course being the Scottish word for an uncooked form of porridge that is mixed with boiling water. The legend was passed down from generation to generation and it was not long before the whisky-honey-oat mixture became a popular drink and was famously used in 1745 by John Stewart, the 1st Earl of Atholl, to spike the drink of a rebel leader, which successfully stopped a Highland rebellion before it was properly started. From then on, the mixture went by the name of Atholl Brose and is much drunk in Scotland. In fact, it is a favourite during the festive period. It is worth perhaps, if you ever get to taste this delicious Scottish drink, holding the glass aloft and toasting that long-dead giant, cruelly slain within his slumber.

Whisky has not only been used to slay giants but the legend of Highlander Tom Campbell tells us that the drink could even be used to keep the devil at bay. It all began many centuries ago when sailor Campbell returned to his hometown of Wigtown, where he took up a job with a blacksmith and settled down with a beautiful local girl. Soon the happy couple had two bairns. Tom was happy and contented with his life and, being a Highlander, he enjoyed a dram or two of the hard stuff. However, it was not all plain sailing, the legend tells us, and disaster struck when a plague hit the town.

One night after finishing work Tom popped into the local tavern and had several quick drinks. He purchased a bottle to take home with him, but before leaving he said to the entire tavern: 'This plague is the work of the devil, but he'll not get the better of me.'

On his way home Tom was stopped by the devil itself suddenly appearing on the road ahead of him.

'I hear you've been laughing at my expense,' the devil said. 'The time has come to pay with your soul.'

'Oh, I was expecting more,' Tom said as he looked the devil up and down. He was of sound courage due to the whisky he'd just drunk and he let out a laugh. He raised his bottle and added, 'Will you take a drink with me?'

The devil looked at Tom with curiosity, turning his head in the way a dog does when it hears a strange sound, but then he snatched the bottle from the man, tore out the cork and started gulping down the strong spirit.

'Save some for me,' Tom jumped forward and snatched the bottle back, dismayed to find that more than half of it had been drunk.

'We will fight for your soul,' the devil said, swaying a little on his feet as the whisky took effect.

Tom nodded.

'We shall fight under strict rules of *cothrom na feinne*,' the devil said. This was a Gaelic phrase that meant 'a fair chance'.

'What if I win?' Tom asked, aware that the devil was more than a little tipsy.

The devil smirked, 'Name your prize.'

'You shall remove this plague.'

'Agreed,' the devil said.

'Then let us have another drink,' Tom said, taking a long gulp and then handing the devil the bottle.

The devil drained the last of the bottle, threw it aside and bellowed: 'We begin.'

They started to wrestle. The devil was much larger and stronger than Tom, but the man took extra strength from the whisky he had drunk, while the devil was unsteady on its feet and, unused to strong whisky, was faltering. They fought on all through the night and were still going as the dawn broke. They continued to wrestle well into the morning, but then devil's foot slipped and Tom took his chance and threw the devil into the air and pinned him down when he landed upon his back. The devil cursed and then vanished in a puff of smoke.

Exhausted, Tom fell to the ground and went into a deep sleep. It was several hours later that the local priest together with Tom's wife came upon him.

As the priest lifted Tom to his feet, he turned to the woman and said, 'It is a double boon, indeed. Your husband is safe.'

'Double boon,' Tom asked groggily, 'What do you mean?'

'The plague lifted this morning,' the priest explained. 'The people are no longer sick and now we find you safe and sound.'

Tom looked first to the priest and then to his wife, and quickly came out with the incredible tale of the bargain he had made with the devil. Of course, very few people believed the tall tale, but all had to admit it was a good way of explaining the empty bottle of whisky at his side, and the fact that he had been out all night. Truth or not, the tale grew in the telling and it soon became a part of Scottish myth and legend.

Chapter 13

How to Taste Whisky

I'd like to start this chapter by making the admission that I shall be ignoring many of the so-called rules that have sprung up surrounding the enjoyment of whisky. In fact, I shall go further and take that rule book, run it through the shredder and add it to the compost heap. Those, and there are still many, who claim that the only way to drink whisky is neat will find no quarter here.

It's an interesting point that these days many consumers who have come to whisky and discovered the wonders of this truly complex drink are taking it long or in a cocktail, and there's nothing at all wrong with that. As respected whisky writer Dave Broom explained in his 2014 work *Whisky: The Manual*, 'The way that most of the new converts to its [whisky] charms, prefer to take the drink is mixed.' And, as stated, there's nothing wrong with that; in fact it should be encouraged. The old rules of how to drink it, what time to drink it, where to drink it, what sex you have to be to drink it, are all hangers on from a different age, a far less-enlightened age. Thankfully, whisky has long since cast aside the image of tweed-suited, monocle-wearing drinkers sitting around the dinner table, a crystal glass in hand, cigar clamped between the teeth, because it's old fashioned, irrelevant even and certainly doesn't reflect the modern way that whisky is consumed. There's precious little tweed in bars and restaurants around the world, unless, of course, I'm in the house. I am quite fond of a good tweed jacket, but in truth I seldom take my whisky totally neat. I tend to prefer it over ice, though there are times, particularly on a warm summer evening, when I will indulge in a whisky mixer. And what a wonderful indulgence that can be. In fact, the only time I take my whisky totally straight is when I'm tasting a whisky for the first time, and even then I tend to add a splash of water after the initial sip or two. Adding a little water to a whisky not only releases trapped flavours but it also sends aromas rushing to the nose. I am aware that any whisky purist reading this will likely now be in a state of apoplexy, but the truth is that whisky is never truly neat, it's already had the addition of water during the proofing stage. Truly neat whisky would likely be unpalatable.

There is actually no right or wrong way to drink whisky, it's all down to the individual. You pay the price and so you should, common sense dictates, drink it the way you like it. If that happens to be neat, then fine. Over ice, equally fine, drowned in cola even or subdued by tonic, or freshened with green tea or ginger ale. Fine, fine, fine – drink it the way you want it, and don't let anyone tell you differently. Although it can't hurt to experiment and towards the end of this book, in the chapter that deals with whisky tastings, I will be looking at the way particular brands change with the addition of certain mixers.

Whisky is all about enjoyment. It's about the delicious anticipation before the caress of the whisky, like a lover's kiss upon the lips. It's the sensory experience as the aroma revitalises olfactory sensations and the wicked sensuousness of the feelings experienced within the mouth. That may seem a little purple, but it's not far from the truth. For whisky truly offers an expansive world of flavours that bring up memories of Christmas cake, of freshly cut straw, forbidden chocolates and fruits long devoured. To paraphrase the great: 'There are places I remember, tastes and sweets long forgotten but of all these sweet sensations, there are none that compare to you.' It's about companionship, music, dance, long conversations, both deep and frivolous. It short, this pure hedonistic pleasure is whatever you want it to be.

That is very much a lesson to be learned – that whisky can be whatever you want it to be. The internet and especially social media have brought this to all our homes. There are scores of YouTube videos in which enthusiastic drinkers offer their own thoughts on the subject. They take the glass to collective snouts and offer up notes of apples, oranges, winter dew and dark decadent chocolate. Even industry experts, master blenders and distillers are not immune to this and they can be seen on their own websites suggesting we should be getting granny's old cakes, forest fires, fig leaves and nylon stockings from the liquid in our glass.

Mind you, has anyone sucked on a pair of old nylon stockings to know what they taste like?

No, on second thoughts forget I asked that question.

Those last few lines are quite facetious, because the complexity of taste in any given whisky can evoke old memories, and the taste can indeed resemble that of say, a long-ago consumed slice of dark chocolate cream cake. Despite the fact that there's no chocolate, nor cream, and certainly no stockings within your whisky, the taste can be there. It truly can. Which is pretty incredible and something quite exclusive to whisky.

To get these associations you need to first relax with your whisky. Sit back in a comfortable chair, rub your thumb over the glass in your hand and then bring it to your nose. Breathe it in – slowly, do not rush – and allow the aromas to waft into your nose. Take it away and then repeat, before bringing the drink to your lips. Hold the liquid in your mouth a moment, chew or swish it around and then slowly swallow. The lingering taste in the back of your throat, called the finish, may even conjure up further flavours. Then add a little water, just the merest drop, and swirl it around the glass. Watch the rivulets, called legs, that slither down the walls of the glass and then take it to the nose and the mouth again. Even different, often more intense flavours, welcome you. That may sound pretentious but it is quite true; the more you experiment the better rewarded you will be. It's the reason the beverage is so popular. It is the reason for the drink having survived the centuries with minimal modifications to the first ever whisky that dripped from the still.

The scientific explanation of how we taste is now well understood, and belies the once popular tongue map that many of us were taught in school, and to some extent is still used to demonstrate the way we experience different flavours today. I've seen the tongue map used at various whisky tastings over the years, but it is a load of old codswallop. The map goes: sweet on the tip of the tongue, salt and sour on the sides and bitter on the back but its accuracy has long been called into question.

The human tongue features hundreds of little bumps called papillae, and each of these contain around six taste buds. We also have these buds on the roof of our mouth and on the epiglottis; that flappy-wappy thing that stops food entering our windpipes. Each of these buds contains microscopic hairs, and it is these hairs that send taste sensations along the nervous system and to the brain. These taste buds renew themselves every so many weeks, but as we age not all of them regenerate. So, children can have around 10,000 taste buds while the average adult has between 2,000 and 5,000, which is why flavours we loved as children are often sickly sweet as adults. However, there is much more to this in the way we taste whisky, and we have to also consider our olfactory senses. Smell can have a marked effect on how we taste, as can the visual aspect of what we are tasting. It all works together, and even our emotional state can add to what we are tasting. The only thing that is certain is that that tongue map we were taught in school is complete and utter tosh, and we are just as capable of tasting sweetness in the back of our mouths as we are in the front.

So how many actual flavours are there? Well, if we consider the five primary tastes to be sweet, sour, bitter, salty and savoury, as the tongue map teaches us, and add to that the variations in intensity of each flavour – scientists say there are ten levels of intensity – then as many as 100,000 flavours are possible. Now if we add to these secondary aspects such as surroundings, mood, temperature, companionship – and yes these do influence the tasting experience – then the range of possible flavours experienced is truly mind boggling. This is why two people can experience tastes quite differently from each other when drinking the same whisky.

A glass of the amber gold from the Cotswolds Distillery. *Cotswold Distillery*

My own tips, for what they are worth, when tasting whisky:

> *When you bring the whisky to the nose, alternate breathing through the mouth and nose to fully awaken all your senses.*
>
> *Swish the whisky around the mouth before swallowing. Engage and occupy all those taste buds.*

If taking notes in a notepad, then be aware of your surroundings and mood and make a note of both.

If the whisky feels too hot, ethanol heavy, then don't be afraid to dilute it with a little water. You could even add ice to bring down the temperature with the glass. Ensure it feels comfortable in your mouth.

Enjoy that finish, that lingering aftertaste, because new taste sensations can come to the fore even after you have swallowed.

Congratulations, you have mastered the art. An entire world of flavour awaits.

Chapter 14

Let's Taste Whisky and Throw in Some Mixers

I promised earlier that this book would contain tasting notes and here they are.

However, these are not the standard tasting notes that can be found in the many other whisky books that are out there, and there are many. Nor is this a selection of what I consider the best whiskies. That would be a book in itself.

Instead, I have limited the tastings to a select few whiskies, with several rules for inclusion. Firstly, they must be easily available and, most importantly, affordable. I see no use in extolling the virtues of a hugely expensive and extremely rare single malt if most of us will never get to experience it. Likewise, the limited editions that have collectors rushing for their credit cards are beyond the brief of this book. These are tasting notes for the everyman, the whisky enthusiast who demands quality on a budget and isn't easily bamboozled by fancy advertising and outlandish claims.

I have also tried to recommend the possible mixers that go well with a particular whisky – ginger ale, cola, soda water, green tea and even coconut water. The latter is fast becoming a popular mixer for whisky and you can blame writer Dave Broom for that. It was he who extolled the mixer's merits in his seminal work *Whisky: The Manual* (Octopus Publishing, 2014). This is the book I would point the reader to if they seek a detailed exploration of consuming whisky as a mixed drink.

So read on and I do hope you try many of the inexpensive options listed over the following pages and are of a mind to experiment with the way you try them. Do try the mixers recommended, and don't make the mistake of thinking that whisky is just another drink when it is so much more than that. It holds a fascination for enthusiasts around the world, and truly is an experience like no other.

Cheers, skål, slainte, bottoms up!

Scotch Blends

Blended Scotch makes up the biggest-selling whisky category in the world, and counts for more than 90% of global sales, with the remainder being made up of everything else. So do the maths, there are more than a hundred malt distilleries in Scotland alone, and none of these could function based on the limited, though growing, share of the whisky market held by single malts. Those distilleries owe their very existence to the blended whisky industry.

It is thanks to the blenders that malt distilleries are able to make a profit by supplying their whisky to the blending houses who blend the malts and single grain whiskies to come up with a consistent product time after time. The art of blending, and it is an art rather than a science, comes only from a deep understanding of the nuances of each single malt and grain whisky, and the qualities that each can bring to a blend.

Bell's

Bell's is one of those whiskies that seems to have always been available; even during the lean years of the 1970s the brand was a common sight in public houses and clubs. Likely it was the whisky your grandfather drunk. There's a reason for that and even if it doesn't have the flashy image of many of the brands often seen alongside it on supermarket shelves, it certainly delivers a flavourful and satisfying experience. Bell's should not be ignored.

The whisky your grandfather drank. *Author's collection*

It delivers a smooth taste, with some honied sweetness, a little wood and overall has a nutty spice about it. The addition of a splash or two of water brings out the fruity characteristics and as a mixer it goes especially well with ginger ale.

Black Bottle

Established in 1879, this used to trade on being one of the smokiest blends available, but over the years the peaty aspect of the drink has been lessened and the modern blend is a mellow sipping whisky with plenty of complexity. There is some smoke but it's no longer front forward and instead we are rewarded with notes of toffee, fruit cake and some cereal.

Smooth, mellow and fruity. *Author's collection*

Adding a little water allows the flavours to linger on the palette and as a mixer it works well with both ginger ale and coconut water.

The Famous Grouse and Smoky Black

Scotland's biggest-selling blended Scotch, and we have two versions here: the regular Grouse with which most whisky drinkers will be familiar and the peated version that used to be called Black Grouse but is now marketed as Smoky Black. The original Grouse is perfectly balanced with marmalade, toffee and dark fruits coming through, while the Smoky Black is a little more complex with the light smoke mingling with dark fruits and peppers. It is not known what whiskies are blended into Grouse but we do know that it contains, among others, some Macallan, Glenrothes and Highland Park.

Both versions work well as a mixer, with the original benefitting from ginger ale, while the latter does best as a smoky cokey; that is to say, mixed with cola.

Scotland's famous bird. *Author's collection*

Grant's Triple Wood

This blend was for years known as Grant's Family Reserve, but in 2018 it had a name change and is now known as Grant's Triple Wood, which plays on the fact that it goes through three maturations in three different

Three times in the wood. *Author's collection*

casks, and therefore takes flavour from each. Virgin oak is first, before being moved into American oak casks and then finished in first fill bourbon casks. It's one of the most complex blends in its price range with a wealth of flavours including a rich dark chocolate on the finish.

As a mixer it goes well with both coconut water and ginger ale, but it is with cola that it really sparkles. Cola brings out a light smoke and allows for the fruity qualities of the Scotch to come through.

Johnnie Walker Red Label

The world's biggest-selling blend and although it is often overshadowed by the Black Label, the red is perhaps the perfect blend

The famous striding man. *Author's collection*

for mixing. While it's the cheapest in the Johnnie Walker range, the quality is there and it is certainly not bland in the taste department. Lemon, ginger, vanilla, fruit and, of course, that signature smoke all leap on to the palate.

Used as a mixer, it is once again ginger ale that offers the best results. Cola tends to overpower the smoky quality that is a signature of the entire Johnnie Walker range, but coconut water does offer a refreshing twist although it tends to dial back that smoke. Green tea surprisingly adds a lot of depth to the drink.

Whyte and Mackay

This is the blend I drink more than any other, but that's only because it is so readily available and fairly priced in my neck of the woods.

Full bodied and very fruity. *Author's collection*

That it offers a great full-bodied taste with strawberry jam, dark fruits and the subtle smoke is a bonus. Whyte and Mackay rocks with the addition of just a little water, but it can certainly serve well as a mixer.

Ginger ale doesn't quite cut it here and subdues much of the native flavours that are an essential part of the whisky, but cola seems to accentuate those flavours, bringing them to the fore and is indeed a hit rather than a miss.

The Mixers

Despite the purists who say that whisky, particularly single malts, should be drunk neat I will say nonsense. Water, just a splash, doesn't ruin a good whisky. In fact, it enhances it and takes away some of the burn that most whiskies possess. Years ago, when I first started visiting whisky bars there was usually a small spigot on the bar so that drinkers could dilute their whisky to their own preference, and if that was not the case then there was more often than not a jug filled with chilled water to be used instead. Whisky should be about pleasure, not pain and why anyone would want to drink a 50% or above ABV whisky neat and then gurn as the burn hits their chest is beyond me. I myself drink a fair number of single malts and if not with the addition of water, then they are taken over ice. Though to be honest most single malts I take with a splash of water, maybe a teaspoon full, while I keep the ice for bourbon and blended whiskies.

Water does wonders for whisky and its addition subdues the burn, as well as releasing flavours. Think of how a welcome summer shower allows the hidden aromas of a sun-baked landscape to come to life. The same is true of whisky and when the ABV of a spirit is more than 20%, which all whiskies are, then the ethanol acts as a roadblock, keeping flavours and aromas back, but when water is added the lights suddenly turn green and flavours and aromas come rushing around that bend.

To take water a step further is to move onto soda water, carbonated water which adds bubbles to our drink and from there we can experiment with adding flavours to our whisky with the use of ginger ale, cola and so on, but never lemonade because that's a step too far. For some reason lemonade doesn't seem to work with whisky. I don't know why. It just doesn't.

Malts

There is a belief that malts don't mix well, and there are also a great many who insist that malts shouldn't be mixed. While the latter may be true of the more expensive malts there is still scope for mixing it up with malts. In fact, some malts absolutely scream to be used as a mixer in their advertising, and a great many actually trade on their suitability as the backbone of a whisky cocktail. All of the whiskies featured over the following pages work wonderfully neat or as the core of a whisky-based mixed drink.

Ben Bracken Islay Single Malt

When I first tasted this whisky, it floored me. I couldn't believe that something so cheap, a supermarket brand, Lidl no less, could be so complex, deliver such a myriad of tastes. Of course, that signature Islay smokiness provides the backbone to the whisky but there is so much more coming through and it certainly doesn't taste like a budget malt.

I discovered the whisky due to the fact that the brand was getting a lot of newspaper coverage, winning several awards along the way and generating column inches in the quality press. I learned a valuable lesson there – my own snobbery centred around supermarket brands

The Ben Bracken ultra-cheap Islay whisky. *Author's collection*

had prevented me from even trying the whisky until influenced by the reports, and when I did it was quite an experience. Heavily peated whiskies are something of a favourite of mine, and this one can stand without shame alongside many far more expensive and well-known peated malts.

Ben Bracken isn't even a distillery, but rather the umbrella branding that Lidl have applied to their range of single malts and it can't be ascertained what distillery provided the malt contained within the bottle. Online whisky detectives seem to favour 3-year-old Bowmore as the liquid used here, but it matters not since wherever the source the quality is there in the bottle.

Undoubtably a young malt, it is nevertheless bursting with the smoky qualities that the Islay region is renowned for. There's thick toffee and peat on the nose, and that smoke carries over to the taste but it's not a one-trick pony and strong fruit notes come through on the finish. As a mixer, why not since it's cheap enough to experiment with. Both cola and ginger ale were a miss for me, but the addition of soda water added a fair amount of complexity and subdued the smoke a little, and coconut water produces a tasty treat.

Dalwhinnie 15-Year-Old

Coming from Scotland's highest-placed distillery, being 326m above sea level, this Highland malt takes its water from Lochan na Doire-uaine and the hardest thing about it is all those pronunciations. The word Dalwhinnie itself is Gaelic for 'meeting place' and perhaps derives from the long-gone days of whisky smugglers and illicit distillers. On the nose it is sweet with a light smoke and this carries over to the taste.

As a mixer you can throw anything at this one. Ginger ale brings out a honey-like quality, while cola enhances the light smoke. Soda water, green tea and coconut water simply do their thing and add their own dimensions but none fuddle the intrinsic qualities of this whisky.

A quality single malt that won't break the bank. *Author's collection*

Glenfiddich 15-Year-Old

The 12-year-old Glenfiddich is the world's biggest-selling Scotch malt whisky, but the 15-year-old is, to my tastes, that much richer on the palate. Those extra three years in the cask provide great dividends to the whisky enthusiast. The whisky is not shy in coming forward and provides tastes of fresh toast smeared thickly with jam, rich chocolate and a citric zest. It's absolutely dripping with flavour and being from the Speyside region is smooth, light and easy to drink.

One of Speyside's finest. *Author's collection*

Used as a mixer it works with both cola and ginger ale, but it's so smooth that it excels with just a little soda water, or even spring water for that matter.

Penderyn Myth

There are many fine whiskies in the Penderyn rage of expressions, but Myth is perhaps one of the most versatile and delights as both a straight drink and a mixer. This is, of course, a Welsh whisky and

The pride of Wales. *Author's collection*

while Penderyn is not the only Welsh distiller it is arguably the best known and certainly boasts the widest distribution. Using water sourced from the Brecon Beacons, Myth provides the taste of tropical fruits, citrus and wood notes.

As a mixer it does really well with ginger ale, while not so good with cola, but it's so light and smooth that perhaps it works best with the addition of soda water.

Smokehead

While most whiskies lean towards traditional designs in their bottles and labels, Smokehead takes a leaf out the craft beer

A modern eye-catching design for this Islay single malt. *Author's collection*

playbook. Tartan, stags and bagpipers are not for this brand but instead it dares to presents its drink as ultra-modern and aims for the hearts and minds of younger whisky enthusiasts. Their smoked glass bottles with a label dominated by a golden skull certainly stands out on the supermarket shelves. The liquid inside the eye-catching bottle is an Islay single malt, heavily peated and full bodied.

When drunk neat the whisky may be a little too smoky for some, but it excels as the base for a mixer. As a smoky cokey it is simply wonderful, and ginger ale tends to bring out a fruity tang among all the smoke. Coconut water also works surprisingly well but both green tea and soda water are simply lost, drowned out by the peat.

Bourbon and Other American Whiskey

American whiskeys are perhaps the most varied in the world, with numerous styles being produced, from rye to bourbon, to single malts and mostly everything else, but the style most associated with the USA is, of course, bourbon: the national drink that can only be made in America.

Buffalo Trace

This is a fine bourbon for its price and can stand proudly alongside far more expensive bottlings. It's got that sweet, oaty quality that is a signature of good bourbon and tastes absolutely superb when taken over ice. Honeyed corn and chocolate are the main notes, but there is a lot of complexity to this drink.

As a mixer it is excellent with ginger ale, but not so much with cola. Coconut water and soda work well here, both bringing out herbal notes in the drink, but green tea falls flat for me.

A fine example of quality bourbon. *Author's collection*

Jack Daniel's Old No. 7

Frank Sinatra's favourite tipple, Jack Daniel's is a bourbon in all but name and it is one of the sweetest whiskeys out there, thanks to being filtered through charcoal prior to being aged in charred oak barrels. It bills itself as a Tennessee whiskey rather than a bourbon, but it does meet the federal standards to be identified as a bourbon. However, because of the extra charcoal filtration, known as the Lincoln County Process, Jack Daniel's insists it is a separate entity to bourbon.

Technically a straight bourbon. *Author's collection*

Whatever one calls it, there is no mistaking that smooth, sweet and mellow taste that has made it America's best-selling whiskey.

As a mixer there is, of course, only one option and that is the iconic Jack and cola, beloved by rebellious rock stars and crooners alike.

Wild Turkey 101

Wild Turkey would most certainly be my desert island bourbon. It's wonderfully flavourful with notes of vanilla, citrus and butterscotch as well as a tantalisingly bitter finish that leaves the taste of cloves in the back of the throat thanks to its generous rye content in the mash bill.

As a mixer it is once again ginger ale that best complements its native flavours and helps bring out a creamy chocolate quality to the drink.

A true desert island bourbon. *Author's collection*

Glossary

ABV or alcohol by volume. This is the amount of alcohol contained within a whisky.

Angel's share. This is a wonderfully romantic term for the amount of whisky that evaporates from the barrel, vanishes into thin air, during the ageing process.

Blended whisky. This is self-explanatory and means that a given whisky is made up of two or more single malts as well as grain whisky. A blended malt is also a blend but made up of only malts and no grain whisky.

Bourbon. All bourbon is whiskey but not all whiskey is bourbon. It must be made up of at least 51% corn and distilled in America.

Casks. Also called barrels or butts, a cask is simply the barrel within which the whisky ages.

Cask finish. This simply means that a matured whisky has been transferred to another barrel for further ageing in a second barrel.

Chill filtration. The process is often used prior to bottling and is where the whisky is chilled down to a low temperature and then passed through a filter to pick up tiny particles, called esters, within the drink. This is done so that the whisky remains crystal clear when water or ice is added to the drink but there are those that claim chill filtration ruins the tastes of a whisky.

Dram. This is a traditional Scottish word to refer to a glass of whisky but does not specify the measure of the drink.

Expression. This simply means the style of any given whisky.

Grain whisky. In Scotland grain whisky is traditionally made from maize corn or wheat.

Malt. Simply means malted barley.

Mash bill. Basically the recipe for a whisky.

NAS. Simply means 'no aged stated' and more and more whiskies are coming out as this, which doesn't mean the quality isn't there but just that the whisky has been aged for the minimum time the rules allow. For instance, Scotch must be aged for a minimum three years, while straight bourbon requires two years.

Nose. There are two meanings here. The first is the aroma, while a secondary meaning is to smell the whisky. For instance, when someone says that they are going to nose the whisky, they are basically saying they are going to smell the drink.

Palate. The flavour profile of a whisky.

Proof. The amount of alcohol a whisky contains. Basically, means the same as ABV.

Rye. Rye whisky must be made from at least 51% rye grain, but that's only in America and even then it's only straight rye, while there any many other styles. Canadian whisky is often called rye whisky for historical reasons and may not even contain any rye at all.

Single cask. A whisky that is solely drawn from one cask and usually bottled at cask strength with no water added for proofing.

Single malt. A whisky blended from various malt whiskies but all from a single distillery, as opposed to a blended malt which is made up of malts from two or more distilleries.

Tasting notes. These give an indication of the flavours found in a whisky, but these are not set in stone since everyone's palate differs.

Triple distillation. A whisky that has been distilled three times. Most Irish whiskies are distilled thrice, while Scottish versions are usually run through the still only twice.

Suggested Reading

There are countless books that touch on whisky and whisky appreciation. Here are just few of my favourites.

Andy the Highlander – *Lochs and Legends* (Harper Collins, 2024)

Billet, Mike – *Peat and Whisky* (Saraband, 2023)

Broom, Dave – *Whisky: The Manual* (Mitchell Beazley, 2014)
– *A Sense of Place* (Mitchell Beazley, 2022)

Bryson, Lew – *Whiskey Master Class* (Harvard Press, 2020)

Jackson, Michael – *Malt Whisky Companion* (DK, 2022)

Maclean, Charles – *MacLean's Miscellany of Whisky* (Little Books, 2006)

Index

Alexander *of Aphrodisias* 4
American War of Independence 64-7
Angel's Share *(movie)* 4

Batman 2
Bean Swaney 52-6
Beatles The 86
Blade Runner *(movie)* 2
Bladder Men 17
Bogart Humphrey 89
Bond James 2, 138-9
Boone Daniel 63
Bourbon County 80
Bowmore 47
Breaking Bad *(TV)* 2
Brackenridge Hugh 75
Broom Dave 160, 166
Bruce Robert the 18
Burns Robert 9, 25, 87
Bushmills 31-2, 40

Campbell Archibald 47
Campbell Daniel 56
Canterbury Tales *(Fiction)* 6
Capone Al 104, 108-12
Carswell Thomas 20
Charles King 121
Chandler Raymond 2
Chester Robert of 5
Churchill Winston 2
Coffrey Aeneas 20, 35-8, 133
Colquhoun Jessie 150-1
Cor Friar John 6
Craig Elijah 80-3
Cromwell Oliver 32
Curtsinger Gilbert 58-61

Doyle Arthur Conan 48
Duffy William 140-4
Dylan Bob 140-4

Easter Uprising 38
Eighteenth Amendment 103
Elizabeth 1st Queen 24

Ferintosh 23-5
Fleming Patrick 87
Forbes Patrick 87

George II King 22
Glenrothes 47

Hall William 20
Hamilton Alexander 70-5
Hemingway Ernest 141
Highland Park 47
Holmes Sherlock 48

Irish Distillers Company 40-1
Irish War of Independence 38

Jabir Al 3, 5
Jacobite Uprising The 24
James IV 6, 24
Jamestown settlement 62-3
Jefferson Thomas 79
Johnny Walker 2, 85
Johnston Robert 73-4
Jura 47

Laverty Paul 92
Lincoln County Process 184
Loach Ken 92

MacKenzie John Dr 28
Mad Men (TV) 2
Malt Tax 18
Makalanga Biawa 43-6
Marlowe Phillip 2
Mathew Theobald 40
May William 20
Merthyr Uprising 121
Metallica 86-7
Militia Act 77
Miller Oliver 77
Moonshine 16, 75
Monkey Shoulder 129-30
Moryson Fynes 7
Muintir Eolais 6-7

Nation Carrie *(Hatchett Granny)* 99-102
Neville John 77-8

Onstwedder John 126
Orange *William of* 24

Passport *the cat* 49-51
Patrick *saint* 14-15
Peaky Blinders (*TV*) 2
Penderyn Dick 122-5
Perry Commodore 116-18
Phillips Thomas 31
Porteous 20-3
Portland *Rum Riot* 97
Prohibition *American* 40, 95-112

Raghnaill Richard Mac 6
Remus George 20
Robert *the Bruce* 18
Robertson George 20
Rolling Stones the 86

Scotch *whisky regulations* 28-9
Shackleton Ernest 144-9
Shaw George Bernard 2
Skalk 8
Sinatra Frank 2, 184

Smith George 28
Spear Jacob 83-4

Taketsuru Masataka 118-19
Talisker 2
Tolbooth Prison 21-3
Tom *the tinker* 76
Towser *the cat* 48-9
Twain Mark 2
Tydfill Merthyr 122

Union Act of 8, 17

Virginia Land Act 64

Wallace William 18
Walgreens 103
Walpole Robert 22
Washinton George 68-79
Wilson Andrew 20
Wilson Cedric 46
Whisky Galore 89-92
Whisky Rebellion 73-9